S0-DYR-923

CHECKING AND COORDINATING ARCHITECTURAL AND ENGINEERING WORKING DRAWINGS

JOHN FREDERICK DUGGAR III

Architect • Atlanta, Georgia

CHECKING AND COORDINATING ARCHITECTURAL AND ENGINEERING WORKING DRAWINGS

McGRAW-HILL BOOK COMPANY

New York St. Louis San Francisco Auckland Bogotá Hamburg Johannesburg London Madrid Mexico Montreal New Delhi Panama Paris São Paulo Singapore Sydney Tokyo Toronto

Library of Congress Cataloging in Publication Data

Duggar, John Frederick III.
Checking and coordinating architectural and engineering working drawings.

1. Architectural drawing. 2. Engineering drawings.
I. Title.
NA2705.D83 1984 720'.28'4 83-14421
ISBN 0-07-018023-7

2 3 4 5 6 7 8 9 0 DOC DOC 8 9 8 7 6 5

ISBN 0-07-018023-7

The editors for this book were Joan Zseleczky and Dorick Byard, the designer was Elliot Epstein, and the production supervisor was Thomas G. Kowalczyk. It was set in Baskerville by Achorn Graphics.

Printed and bound by R. R. Donnelley & Sons Company.

Dedicated to my family and to friends in the architectural and engineering professions.

CONTENTS

PREFACE

The checking and coordinating of architectural and engineering drawings is one of the most exacting and crucial tasks in an architectural office. Drawings which are poorly checked and coordinated can be a source of many problems, not only during the production of the drawings, but even more so during the bidding and construction phases of a project. These problems can be serious and costly, which makes it imperative for the drawings to be checked and coordinated very carefully.

This book presents a system of graphic techniques for checking and coordinating architectural and engineering drawings which, if carefully applied, can virtually eliminate errors, omissions, mistakes, and duplications. The graphic techniques provide positive assurance that no item on the drawings will be missed in the checking, and that each item will be coordinated with all divisions of the work to which it might relate.

In addition, the techniques described in this book provide a unique system for communicating with the various persons involved in a project. Techniques for getting questions answered and for getting revisions made are based on graphic methods which make the necessary communications precise and efficient, minimizing the risk of any misunderstanding.

The chapters which follow describe how the various techniques used in this system are applied, and will further discuss the advantages of the system.

The ability to check and coordinate drawings depends on experience and natural ability. This book uses the pronouns *he* and *his,* not because of any bias, but due to the demands of the language. Use of *he/she* and *his/hers* was felt to be awkward and distracting. Other expedients can lead to language which is imprecise or confusing, which is obviously not acceptable when new or complicated technical procedures are being described.

Improving the ability to check and coordinate drawings is the primary purpose of this book; any impediment to achieving that purpose would not serve the best interests of anyone. Thus it is hoped that all who use the book will realize why certain usages have been retained.

John Frederick Duggar III

ABOUT THE AUTHOR

John Frederick Duggar III, a former instructor in Reinforced Concrete Design and in Design of Steel Buildings, has worked in both small and large architectural offices, from which he has gained broad experience in all phases of architectural practice. Several years of responsibility for the final checking and coordinating of architectural and engineering drawings showed Mr. Duggar the need for more systematic techniques in that area. This book is the result of the development of these more precise techniques. Mr. Duggar received his Bachelor of Architecture degree from Alabama Polytechnic Institute (now Auburn University), and maintains a practice in Atlanta, Georgia.

CHECKING AND COORDINATING ARCHITECTURAL AND ENGINEERING WORKING DRAWINGS

1 INTRODUCTION

1-1 GRAPHIC CHECKING SYSTEM

A basic concept of this system of checking is the utilization of a variety of techniques which provide a complete graphic record of the status of the checking at all times. This graphic record means that the checker does not have to rely on his memory as to what has been checked, what has not been checked, what his comments and questions were, what additional details might have been needed, what sketches he made regarding various items, what questions he asked other people, and what their responses were. In addition, the graphic record is kept in a logical and accessible manner which contributes in many ways to the usefulness and efficiency of the checking system. Also, using this system, the checker employs techniques which enable him to communicate with others graphically, yet retain a full record of his communications (his marks on checkprints, comments, sketches, questions, etc.). This means that further checking can continue while revisions resulting from earlier checking are being made.

Some of the graphic techniques described below allow the checker to check incomplete sheets or incomplete sets of drawings, and to check additional work later without having to go back and recheck all of the earlier drawings. Furthermore, the checker may selectively check important items first and leave lesser items for later checking. He may also check any desired sheet in any division of the work, which aids in the efficient utilization of the personnel making the revisions. These graphic techniques and their use in checking, coordinating, and communicating will be described in detail in succeeding chapters.

1-2 BASIC ASSUMPTIONS

This presentation assumes a more or less typical work situation, with separate architectural, structural, HVAC (heating, ventilating, and air-conditioning), plumbing, and electrical consulting engineers. In addition, it is assumed that the project is in the working-drawing phase and that it is large enough to require a separate checker, one or more job captains, and several drafters who perform the revisions of the drawings according to notations made by the checker.

Even if a project were larger or smaller than this typical situation, the same basic concepts would still apply. For instance, if a project were very small and only one person were involved, there would still be a need to check the drawings in the same way. In a word, any architectural or engineering drawings, regardless of the size of the project they represent, may be efficiently checked by use of the techniques described in the following chapters.

1-3 OVERLAY TECHNIQUE FOR CHECKING DRAWINGS IN PROGRESS

A unique overlay technique is described in Chapter 16, by means of which the checker may check successively more and more complete prints of drawings in progress, yet does not have to repeat all of the checking each time a new print is made. While this overlay technique is a major aspect of this system of checking and would probably be used on most projects, it is not described until Chapter 16 because earlier chapters are devoted to describing more basic concepts. In the earlier chapters, it is much simpler to describe the checker's techniques used in checking items directly on the check-prints. Then, after those techniques have been presented, they can be readily adapted to the technique of checking on an overlay attached to a check-print.

Remark: The overlay techniques described in this book (including the one referred to above) are not to be confused with the reprographic overlays used in systems drafting. Totally different principles are involved in the overlay techniques presented in this book, as will be apparent from the detailed descriptions in later chapters.

1-4 CHECKING AND COORDINATING DRAWINGS IN CONSULTING ENGINEERING OFFICES

The techniques for checking and coordinating drawings presented in the following chapters are described from the point of view of the checker in an architectural office, but the techniques are equally useful in checking and coordinating drawings in structural, HVAC, plumbing, and electrical engineering offices.

Even on projects where the checker in the architectural office takes responsibility for the overall checking and coordinating, the various engineering consultants could still very profitably utilize many of these techniques in checking and coordinating within their own offices and with other consultants. The result would be engineering drawings which were well coordinated and mistake-free before they were reviewed by the architectural checker. This would minimize the confusion and loss of time caused by having to make excessive revisions of the drawings, and would be beneficial to the architectural and the consulting engineering offices alike.

1-5 CHECKING AND COORDINATING BY PROJECT MANAGERS, JOB CAPTAINS, AND OTHERS

Many of the techniques for checking and coordinating which are presented in the following chapters for use by the overall checker, could also be used to advantage by project managers, job captains, and others. If the drawings were checked and coordinated by project managers, etc., in the early stages of a project, it would aid the overall checker greatly. The project managers and job captains, who are closer to the actual production of the drawings, would be able to resolve many problems at an early stage, before ramifications developed that would be difficult to correct later.

However, it is still imperative to have one person do the overall checking and coordinating. A person who has not been too closely involved in the production of the drawings should be considered for the job of overall checker, particularly for the final check of the finished drawings. Such a person might have a fresh viewpoint, which would enable him to see things on the drawings which might be missed by someone more familiar with the drawings.

1-6 CHECKING AND COORDINATING DRAWINGS USED IN OTHER INDUSTRIES

The techniques for checking and coordinating drawings which are presented in this book can be adapted for use in many areas of engineering and industry other than the building industry. Anywhere drawings are involved and precise communications are essential, a need exists for systematic and efficient procedures for checking, coordinating, and communicating.

Where terminology for other industries needs to be substituted, persons knowledgeable in those industries can use the basic concepts of this book and adapt the terminology as required.

1-7 CHECKING DRAWINGS PRODUCED BY SYSTEMS DRAFTING

In offices which use **systems drafting** (reprographic overlays), these checking and coordinating techniques are completely applicable. Despite the "self-coordinating" aspect of using reprographic overlays of plans, errors can still arise; and obviously all of the elevations, sections, details, and schedules need to be checked, no matter what processes are used to produce the floor-plan drawings.

1-8 CHECKING COMPUTER-AIDED DRAFTING

In offices which use computers in their drafting procedures, these techniques are also completely applicable. As with systems drafting, all drawings need to be checked, no matter what processes are used to produce them.

2 MASTER CHECK-PRINTS

2-1 MASTER CHECK-PRINTS

This system of checking utilizes a set of check-prints which are called the **master check-prints**. They are called this because the checker marks a system of notations on them in colored pencils, and these check-prints and the checker's notations are then used as "masters" in communicating with others. This communication is accomplished by a technique in which certain of the notations are forwarded, copied that is, onto duplicate check-prints and then transmitted to the proper parties. Communication by means of marked-up **duplicate check-prints** allows the checker to communicate with others, yet retain a complete set of master check-prints at all times.

The master check-prints should be carefully kept in proper order and held together in a set with one or more spring clamps. Within the overall set, the prints should be arranged in numerical order according to their sheet numbers (if no sheet numbers have been assigned, the checker should assign temporary sheet numbers). As successive prints are made from the same drawing, the checker should include them in the set of master check-prints. All prints from the same drawing are grouped together. Unlike the normal numerical order of the overall set, the sheets in the groups of prints from the same drawing should be arranged in reverse chronological order, with the latest prints on top.

2-2 DUPLICATE CHECK-PRINTS

Duplicates are made from the checker's master check-prints so that the checker may forward desired information to others. This will be described further in Chapter 11.

2-3 DATING, DESIGNATING, AND ISSUE-NUMBERING OF MASTER CHECK-PRINTS AND DUPLICATE CHECK-PRINTS

The checker should mark each check-print of a drawing with the date it was received and with an issue-number showing its proper chronological position. The checker should use abbreviations in designating the various prints. The first check-print of a drawing should be designated MCP #1 (master check-print #1), and the duplicate check-print should be designated DCP #1 (duplicate check-print #1). In like manner, the second check-print of that same drawing (made after further work has been done on the drawing) should be designated MCP #2 (master check-print #2), the duplicate should be designated DCP #2 (duplicate check-print #2), and so on with subsequent check-prints.

The designations and issue-numbers MCP #1, MCP #2, DCP #1, DCP #2, etc., should not be confused with the normal sheet numbers of the drawings; the designations MCP #1, DCP #1, etc., are additional to the normal sheet numbers and are added in order to differentiate between the master check-prints and the duplicate check-prints, and also to establish the chronological order of the check-prints.

It is important to date and issue-number check-prints so that the checker and other personnel are able to positively identify all check-prints and their chronological order. This prevents the confusion which often arises when one person is referring to a certain issue of a drawing and another person is referring to another issue of the same drawing.

These dates, designations, and chronological issue-numbers should be made in dark-green pencil (such as Eberhard-Faber Colorbrite Green #2128) and placed on the check-prints near the title block. The reason for using a dark-green pencil will be discussed in Chapter 3.

2-4 DIAZO BLUELINE PRINTS

All master check-prints and duplicate check-prints, and any other prints referred to in this book, should be diazo blueline prints.

3 CHECKER'S COLOR CODE

3-1 CHECKER'S COLOR CODE

A second basic concept of this system of checking is the use of colored pencils on diazo blueline check-prints to graphically indicate certain things to the checker by means of a simple color code. This **checker's color code** enables the checker to know exactly which items on a check-print have been checked and which have not been checked (as well as other information, as described below). The color code is also used to apply a system of notations on the check-prints. These notations serve as graphic signals which aid the checker in various ways (see Chapter 4).

3-2 COLORS

The basic colors employed in this color code are light green, red, and yellow. The light green and the yellow must be smooth and transparent so that the blue lines, notes, etc., on the check-prints may be read underneath an application of those colors. The need for transparency is the reason for using a light green rather than a darker green.

Eagle Turquoise #2376 colored lead, for use in a lead holder, is recommended for both the light green and yellow colors. Other makes are available, but Eagle Turquoise products have been found to be superior in smoothness and transparency.

The red color must be very intense, to minimize the chance that any red notations placed on a check-print will be missed in the course of the checking. Eberhard-Faber's Colorbrite Medium Red #2126 is recommended for this red, because it is quite intense and has the desirable qualities of being soft and easy to apply (as opposed to some other colored pencils which are hard, scratchy, and brittle).

3-3 ERASABILITY

The main reason colored pencils are recommended rather than felt-tip markers or ballpoint pens is that they are erasable in case of a mistake. For erasing, the Faber-Castell #1962 Auto Magic-Rub vinyl eraser is recommended (the eraser is held in a handy "chuck-type" plastic holder). As an alternative, the A. W. Faber #1960 Peel-Off Magic-Rub vinyl eraser is recommended. Still another alternative would be the Faber-Castell #75 green pencil eraser (or other soft eraser) for use in an electric eraser.

3-4 SIGNIFICANCE OF COLORS

The significance of the various colors and combinations of colors is described in the following articles.

3-5 LIGHT GREEN

A light-green tone is applied on a check-print over each item or element of the drawings (or around the outline of large items) which the checker reviews and *tentatively* accepts as being correct. Examples of the items which have green applied to them (or around their outlines) are the configurations of sections, details, fixtures, and equipment. Also, all lettering and numbers—such as notes, space names and numbers, door and window numbers, titles of all kinds, and schedules—have green applied to them when they are tentatively accepted.

The checker's acceptance of an item is tentative in that, even though it appears correct to the checker as of that moment, it is still subject to further review as it is cross-checked against other points on the drawings.

Light green is also applied in the same manner to each item on the check-print which is *not* accepted as being correct as it appears, or which needs some revision, or which needs to be cross-checked at other points on the drawings; but the light green is *not applied until after the checker makes notations in red which ensure that the item will be properly cross-checked, revised, or corrected.* The light green in this case merely indicates that the item has been reviewed and accepted *provisionally*, and that further action is required, as indicated by the red notations.

Remark: The checker should be very careful to apply all the red notations needed for an item before he applies the light-green tone over it; in this way the danger that the checker will forget to apply all of the necessary red notations is minimized. (Also, when the checker applies light green over the item, he must be careful not to apply light green over any of the red notations, as these notations must remain as reminders until they have been taken into account.)

The general rule, then, regarding the light-green color on a check-print is that, unless red notations accompany the light green or unless subsequent cross-checking indicates a revision of an item is needed, the light green indicates that no further action is required.

Remark: It is reemphasized here that the light green pencil tone should be smoothly and lightly applied, so that the tone remains transparent enough that the print can still be easily read.

3-6 RED

Red is used for marking notations on or near each item on a check-print which the checker feels requires some action. (A system of notations for this purpose is described in Chapter 4.) Red is also used to make a direct indication on the check-print of any desired change that is very simple; this direct indication method is described more fully in Chapter 4.

3-7 LIGHT GREEN OVER RED

A light-green tone is applied over a notation in red on a check-print to indicate that the checker has taken action on that red notation, and therefore that particular notation needs no further attention.

The effect of the light green over the red is to neutralize the red and give a gray-green tone, which, because it is a tone of green, is appropriate to indicate that the notation is of no further concern.

3-8 YELLOW OVER RED (YELLOW/RED)

A yellow tone is applied over a notation in red on a check-print to signify that the notation has been copied onto a duplicate check-print for transmittal to other personnel, such as an engineer or drafter (see Chapter 11). This yellow tone should be applied over each red notation as soon as the notation has been copied onto the duplicate check-print, in order to ensure that all the notations will be transmitted.

In addition, placing the yellow tone over the red notations which have been copied differentiates those notations from notations which were added later. Thus, the checker can tell at a glance which notations have been previously copied and transmitted, and which notations were subsequently added and have not yet been copied and transmitted.

This yellow/red tone also serves as a caution signal to remind the checker to eventually verify that engineers or drafters have made revisions in accord with the red notations which are underneath the yellow tones.

3-9 LIGHT GREEN OVER YELLOW/RED

Light green is applied by the checker over the yellow/red tone (described above) after he has verified that revisions have been made in accord with his notations in red.

The light-green tone over yellow/red changes the yellow/red to a tone of green, which is appropriate to indicate that proper action has been taken and that the red notation and the formerly yellow tone (now both a tone of green) require no further attention.

3-10 DARK GREEN

A green which is adequately dark to differentiate it from the light-green tone is used by the checker for applying titles on prints and overlays, for designating master check-prints and duplicate check-prints, for indicating the order in which check-prints are issued, and for indicating dates on check-prints (see Chapter 2).

In addition, dark green could be used for indicating any other information on the drawings which the checker wishes to record for future reference; examples are cumulative dimensions used in checking dimensions, a manufacturer's number, or other such items.

Dark green is appropriate for the above uses because its significance is easily included mentally in the same category as the light green, which is to say that it requires no further action on the part of the checker.

Eberhard-Faber's Colorbrite Green #2128 is recommended for this dark-green color. Like the Colorbrite Red described above, the Colorbrite Green is quite intense, and is soft and easy to apply.

3-11 MISCELLANEOUS COLORS

A number of other colors are used for various purposes described in succeeding chapters.

3-12 EXAMPLES OF THE CHECKER'S COLOR CODE

Plate I shows examples of the basic colors of the checker's color code applied to a hypothetical floor plan.

4 CHECKER'S NOTATIONS

4-1 CHECKER'S NOTATIONS

Another basic concept of this system of checking is the checker's use of very concise notations which are placed on the check-prints in red pencil. The notations provide a variety of graphic signals that the checker utilizes in a number of ways. These **checker's notations** and how they are used are described in detail in this chapter.

4-2 ADVANTAGES OF THE CHECKER'S NOTATIONS

The checker's notations are essentially a form of shorthand which enable the checker to indicate a multitude of signals in a minimum of space on the check-prints.

Some of the notations applied on the check-prints use a numbering system which ties that type of notation to a related comment, sketch, or other entry which the checker has recorded in a loose-leaf three-ring binder called the checker's manual (to be described below). Use of this system ensures that the drawings will not be cluttered up with the conglomeration of comments, sketches, etc., which are usually seen on check-prints. Extensive comments and sketches (recorded in the checker's manual) are represented on the check-prints by very small notations.

The notations are also useful in that they can indicate that a comment or a sketch applies at any number of points on the check-prints, eliminating the need for repeating the comment or sketch at all of those points. Merely repeating the notation on the check-prints has the effect of repeating the comment or the sketch which relates to that notation. This avoids the task of laboriously copying comments (or making sketches) over and over at different points on the check-prints. In addition, use of the checker's notations greatly facilitates the checker's communications with others (see Chapter 11).

4-3 EXAMPLES OF CHECKER'S NOTATIONS

The following articles present examples of each of the checker's notations, along with a brief commentary regarding each notation. Detailed descriptions of how the checker uses the checker's notations will be given in subsequent chapters.

4-4 SINGLE-LETTER ALPHABETICAL NOTATIONS

Items marked with the following notations are to be cross-checked by the checker at other points on the drawings as indicated.

B Blow-up (check item against blow-up [larger-scale drawing])

C Ceiling plan (check item against ceiling plan; C1—check item against first-floor ceiling plan; C2—check item against second-floor ceiling plan, etc.)

D Door schedule (check item against door schedule)

E Elevations (check item against elevations)

F Finish schedule (check item against finish schedule)

G Grading plan (check item against grading plan or site plan)

P Plan (check item against plan; P1—check item against first-floor plan, P2—check item against second-floor plan, etc.)

R Roof plan (check item against roof plan)

S Sections (check item against sections)

4-5 TWO-LETTER ALPHABETICAL NOTATIONS (PLUS A NUMBER)

Each of the following notations serves to relate an item on a check-print to a like-numbered comment, sketch, or other entry which the checker records in the loose-leaf checker's manual. (How the checker's notations and the related entries in the checker's manual are used will be described in chapters which follow.) The numbers in the examples below are arbitrary, and are shown for the purpose of illustration.

AC32 Architectural comment number 32. This notation, placed on an architectural master check-print, indicates that the checker has recorded a like-numbered comment in the Architectural Comments section of the checker's manual.

AS26 Architectural sketch number 26. This notation, placed on an architectural master check-print, indicates that the checker has made a like-numbered sketch in the Architectural Sketches section of the checker's manual.

AD12 Architectural detail needed number 12. This notation, placed on an architectural master check-print, indicates that the checker has recorded a like-numbered entry listing a detail needed in the Architectural Details Needed section of the checker's manual.

CR9 Checker's reminder number 9. This notation, placed on any master check-print (architectural, structural, HVAC, plumbing, or electrical), indicates that the checker has recorded a like-numbered miscellaneous reminder to himself in the Checker's Reminders section of the checker's manual.

NO23 Notation number 23. This notation, placed on any master check-print (architectural, structural, HVAC, plumbing, or electrical), indicates that the checker has recorded a like-numbered entry of supplemental notations in the Notations section of the checker's manual.

Remark: Checker's notations relating to comments, sketches, and details needed in the structural, HVAC, plumbing, and electrical drawings are similar to the notations relating to comments, sketches, and details needed in the architectural drawings, as described above (at AC, AS, and AD).

Thus, checker's notations on the structural, HVAC, plumbing, or electrical master check-prints follow the above descriptions verbatim, except that wherever the word "architectural" appears above (at AC, AS, and AD), the appropriate word "structural," "HVAC," "plumbing," or "electrical" is substituted. Also, wherever the abbreviation "A" appears above (at AC, AS, and AD), the appropriate abbreviation "S," "H," "P," or "E" is substituted.

As an example, the checker's notation SC32, placed on a structural master check-print, indicates that the checker has recorded a like-numbered comment in the Structural Comments section of the checker's manual.

4-6 MISCELLANEOUS SYMBOLS

□ Item thus marked needs further action by checker. Action is not specified; notation is merely a signal to remind checker that item needs further attention.

× Item thus marked to be removed from drawing.

↰ Place thus marked needs section cut.

4-7 HIEROGLYPHIC NOTATIONS

Items marked with the following notations are to be cross-checked by the checker at other points on the drawings as indicated.

○ Architectural cross-check (checker will check item against architectural drawings)

– Structural cross-check (checker will check item against structural drawings)

\ HVAC cross-check (checker will check item against HVAC drawings)

| Plumbing cross-check (checker will check item against plumbing drawings)

/ Electrical cross-check (checker will check item against electrical drawings)

P
H \ | / E
S — O
A

Figure 4-1

These hieroglyphic notations may be more easily remembered if they are pictured as shown in Figure 4-1. It will be noted that in this form, if the indication for architectural were taken as a center, the indications for structural, HVAC, plumbing, and electrical would be arranged in a clockwise manner in the same order normally used in binding a set of architectural and engineering drawings. In addition, the alphabetical indications of the various divisions of the work, taken in the order shown, spell the acronym "ASHPE," which can be easily memorized.

The five hieroglyphic notations may be placed in any convenient arrangement on the check-print, on or near the item to which they refer. They may be either clustered or scattered around the item, depending on the space available on the check-print. Examples of a few of the many arrangements which might be employed to fit available space on a check-print are shown in Figure 4-2.

Figure 4-2

In a short time it becomes second nature to recognize the significance of these hieroglyphic notations, no matter how they are arranged. This is the key to their use: a check-print marked with any number of these various hieroglyphics (and other notations) can be scanned and all the indications of a particular hieroglyphic can be rapidly picked out. This feature is important in cross-checking an item on a drawing against drawings of other divisions of the work (see Chapter 10).

4-8 DIRECT INDICATION

In addition to using the checker's notations, the checker can make a direct indication on the check-prints of any desired directive that is very simple. A direct indication would be appropriate to show that an item needs to be moved, revised slightly, or added to a drawing. Very brief comments or questions may also be written directly on the check-prints. However, if a directive is complex at all, it should not be handled by direct indication but rather indicated by providing an appropriate checker's notation and by making a related comment or sketch in the checker's manual.

4-9 REMOTE CHECKER'S NOTATIONS

If necessary, a checker's notation may be somewhat remote from the item it refers to on the master check-print. In this event, the checker uses an arrow to connect the notation with the item.

4-10 COMBINED CHECKER'S NOTATIONS

When a number of different checker's notations are applied in a group on a master check-print, the checker should use a single parenthesis between each notation to avoid confusion. For example, assume that a series of notations are placed in a group on a master check-print. They would appear as follows: AC12)AS22)AD7)C2)C3)B)R —\/. (The meaning of each of these checker's notations is described in preceding articles.)

4-11 FURTHER EXAMPLES OF CHECKER'S NOTATIONS

Plate II shows representative examples of checker's notations applied to a hypothetical floor plan.

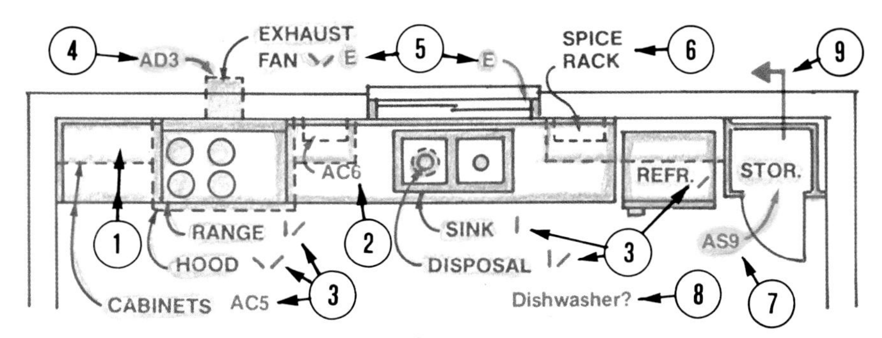

PLATE I
CHECKER'S COLOR CODE
(SEE CHAPTER 3)

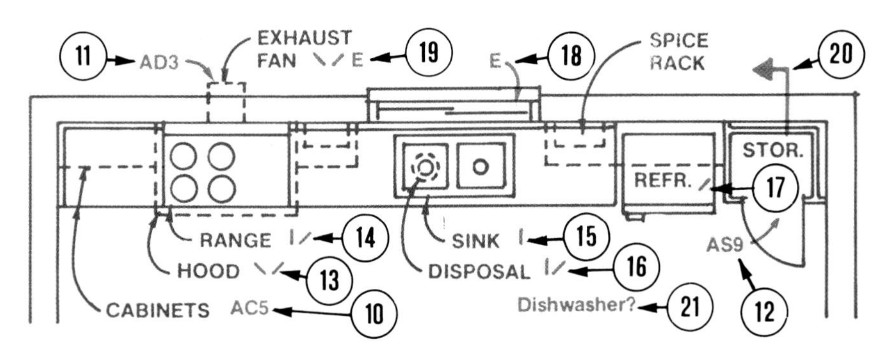

PLATE II
CHECKER'S NOTATIONS
(SEE CHAPTER 4)

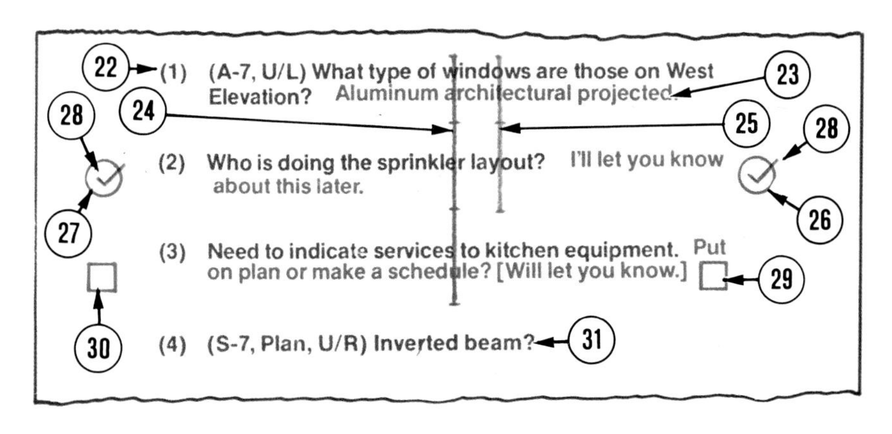

PLATE III
PAGE IN A QUESTION SECTION
(SEE CHAPTER 15)

NOTE: SEE PAGES 19 THROUGH 23 FOR KEY TO MEANING OF NUMBERS IN CIRCLES

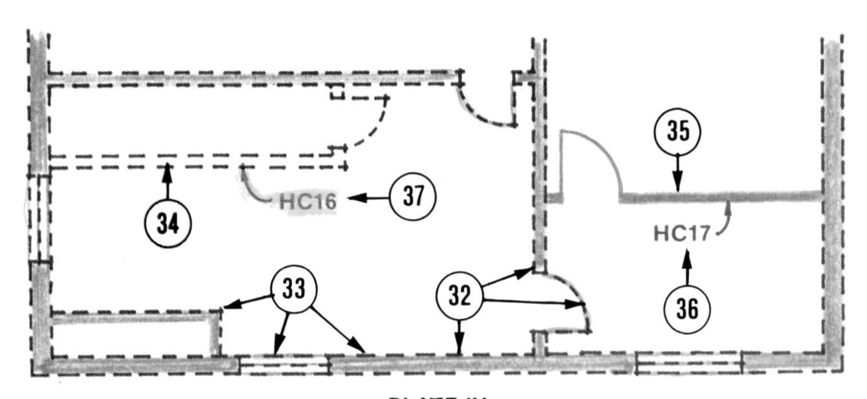

PLATE IV
FLOOR PLAN OVERLAY
(SEE CHAPTER 17)

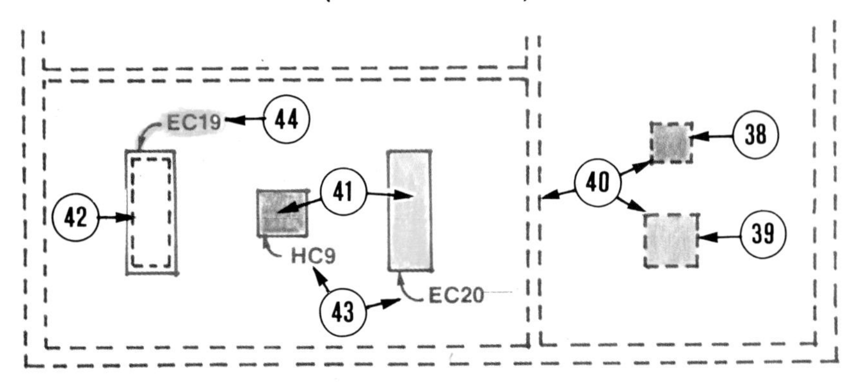

PLATE V
CEILING PLAN OVERLAY
(SEE CHAPTER 18)

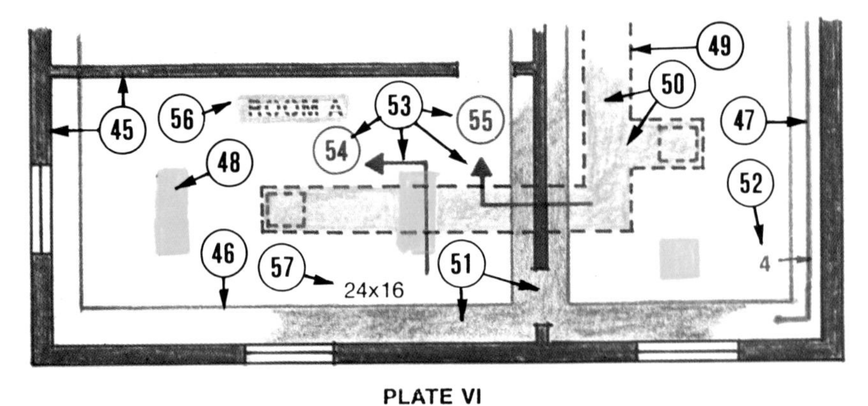

PLATE VI
ARCHITECTURAL AND ENGINEERING OVERLAY
(SEE CHAPTER 19)

NOTE: SEE PAGES 19 THROUGH 23 FOR KEY
TO MEANING OF NUMBERS IN CIRCLES

SECTIONS IN THE CHECKER'S MANUAL

Index Tab	Section Title	Related Notation on Check-Print	Contents of Section
AC	Arch. Comments	AC & number on arch. check-print	Checker's comments regarding items on architectural drawings
AS	Arch. Sketches	AS & number on arch. check-print	Checker's sketches regarding items on architectural drawings
AD	Arch. Details Needed	AD & number on arch. check-print	Checker's list of additional architectural details needed
SC	Struct. Comments	SC & number on struct. check-print	Checker's comments regarding items on structural drawings
SS	Struct. Sketches	SS & number on struct. check-print	Checker's sketches regarding items on structural drawings
SD	Struct. Details Needed	SD & number on struct. check-print	Checker's list of additional structural details needed
HC	HVAC Comments	HC & number on HVAC check-print	Checker's comments regarding items on HVAC drawings
HS	HVAC Sketches	HS & number on HVAC check-print	Checker's sketches regarding items on HVAC drawings
HD	HVAC Details Needed	HD & number on HVAC check-print	Checker's list of additional HVAC details needed
PC	Plumbing Comments	PC & number on plumbing check-print	Checker's comments regarding items on plumbing drawings
PS	Plumbing Sketches	PS & number on plumbing check-print	Checker's sketches regarding items on plumbing drawings
PD	Plumbing Details Needed	PD & number on plumbing check-print	Checker's list of additional plumbing details needed
EC	Electrical Comments	EC & number on electr. check-print	Checker's comments regarding items on electrical drawings
ES	Electrical Sketches	ES & number on electr. check-print	Checker's sketches regarding items on electrical drawings
ED	Electrical Details Needed	ED & number on electr. check-print	Checker's list of additional electrical details needed

PLATE VII

(CONTINUED ON NEXT PAGE)

SECTIONS IN THE CHECKER'S MANUAL (CONTINUED)

Index Tab	Section Title	Related Notation on Check-Print	Contents of Section
CR	Checker's Reminders	CR & number on any check-print	Checker's reminders to himself of actions he must take regarding items on check-prints (or any other items)
NO	Notations	NO & number on any check-print	Checker's supplementary notations which were not put on check-print due to lack of space
AQS	Arch. Quest. Subsections-Index	None	Index of Architectural Question Subsections
AQ1	Arch. Quest. Subsection 1	None	Checker's questions directed to member of architectural personnel designated AQ1
SQS	Struct. Quest. Subsections-Index	None	Index of Structural Question Subsections
SQ1	Struct. Quest. Subsection 1	None	Checker's questions directed to member of structural personnel designated SQ1
HQS	HVAC Quest. Subsections-Index	None	Index of HVAC Question Subsections
HQ1	HVAC Quest. Subsection 1	None	Checker's questions directed to member of HVAC personnel designated HQ1
PQS	Plumb. Quest. Subsections-Index	None	Index of Plumbing Question Subsections
PQ1	Plumb. Quest. Subsection 1	None	Checker's questions directed to member of plumbing personnel designated PQ1
EQS	Elect. Quest. Subsections-Index	None	Index of Electrical Question Subsections
EQ1	Elect. Quest. Subsection 1	None	Checker's questions directed to member of electrical personnel designated EQ1
MQS	Misc. Quest. Subsections-Index	None	Index of Miscellaneous Question Subsections
MQ1	Misc. Quest. Subsection 1	None	Checker's questions directed to person designated MQ1

PLATE VII (CONTINUATION)

(CONTINUED FROM PRECEDING PAGE)

KEY TO MEANINGS OF NUMBERS IN CIRCLES ON PLATES I TO VI

1. Light-green tone over any item (or around outline of item) indicates that item is tentatively accepted, subject to further cross-checking.
2. Red is used to indicate various checker's notations, with meanings as described in Chapter 4 (see also Plate II).
3. Light-green tone *plus* red notation indicates item is provisionally accepted, subject to any action required by red notation.
4. Yellow tone over a red notation indicates that notation has been copied onto duplicate check-print for transmittal to others.
5. Light-green tone over red notation indicates that notation has been taken into account and requires no further attention.
6. No light-green tone over an item indicates that item has not been checked.
7. Light-green tone over yellow tone over red notation indicates notation has been transmitted (per Item 4, above) and has been acted upon by drafter or others, and checker has verified that action taken was in accord with notation.
8. Red is used for direct indication of brief comments and questions.
9. Red is used for various symbols, with meanings as described in Chapter 4.
10. Checker has recorded a comment regarding cabinets; comment is entry number (5), recorded in Architectural Comments section of checker's manual.
11. Checker has recorded a listing of details needed regarding exhaust fan; listing is entry number (3), recorded in Architectural Details Needed section of checker's manual.
12. Checker has made a sketch regarding Storage; sketch is entry number (9), recorded in Architectural Sketches section of checker's manual.
13. Checker will cross-check hood against HVAC and electrical drawings.
14. Checker will cross-check range against plumbing and electrical drawings.
15. Checker will cross-check sink against plumbing drawings.

16. Checker will cross-check disposal unit against plumbing and electrical drawings.

17. Checker will cross-check refrigerator against electrical drawings.

18. Checker will cross-check window against building elevation drawings.

19. Checker will cross-check exhaust fan against HVAC, electrical, and building elevation drawings.

20. Section needs to be cut here.

21. Direct indication of a question. (Checker indicates very brief comments and questions directly on check-print; no entry is made in the checker's manual.)

22. Questions entered in checker's manual, recorded with blue or black ballpoint pen.

23. Answers to questions, recorded with red ballpoint pen.

24. Vertical line drawn by the checker through each question (and answer) when question has been answered. Line is drawn with red ballpoint pen.

25. Vertical line drawn by the checker through each question (and answer), when checker has taken action in accordance with the answer. Line is drawn with green pencil, such as Colorbrite Green #2128.

26. Red circle indicates that checker is waiting for further action by another person. Circle is drawn with red ballpoint pen.

27. Red circle is also shown in border to "flag" that checker is waiting for further action by another person. Circle is drawn with red ballpoint pen.

28. Green check mark placed by checker in red circle indicates that other person has taken necessary action.

29. Red square (plus red brackets) indicates a reminder which checker has recorded of some further action he needs to take. Square and brackets are indicated with red ballpoint pen. (After checker takes necessary action he places a green check mark in the red square to indicate that no further action is required.)

30. Red square is also shown in border to "flag" that checker must take some further action. (After checker takes necessary action he places a green check mark in this red square to indicate that no further action is required.)

31. Question which still needs to be asked of someone is indicated by absence of vertical red line through question.

32. Dotted lines indicate outlines of walls, partitions, doors, windows, equipment, etc., on print of architectural floor plan (below the overlay).

33. Colorbrite Orange #2122 applied to the overlay over print of architectural floor plan; orange is applied to walls, partitions, doors, windows, equipment, etc. (or around the perimeter of large items of equipment).

34. Partition indicated on print of HVAC, plumbing, or electrical plan (below the overlay), but no orange color appears on overlay above (meaning that partition does not appear on architectural plan); indicates a problem which the checker needs to clear up.

35. Orange color on the overlay (representing a partition on architectural plan), but no indication appears on print of HVAC, plumbing, or electrical plan (below the overlay); indicates a problem which the checker needs to clear up.

36. Checker's notation in Colorbrite Medium Red #2126 applied on the overlay; related entry has been made in the checker's manual.

37. Same as 36, except that yellow added over red indicates that red notation has been copied onto duplicate check-print for communication to others. After problem has been cleared up, checker should erase this yellow/red notation and correct the floor plan overlay for further use.

38. Orange plus light green; Colorbrite Orange #2122 applied on the overlay over print of HVAC plan, indicating ceiling diffusers, grilles, etc.; then overlay is transferred to print of architectural ceiling plan and Eagle Turquoise Green is applied over orange where orange agrees with ceiling plan.

39. Yellow plus light green; Eagle Turquoise Yellow applied to the overlay over print of electrical plan, indicating lights at ceiling; then overlay is transferred to print of architectural ceiling plan and Eagle Turquoise Green is applied over yellow where yellow agees with ceiling plan.

40. Dotted lines indicate outlines of walls, partitions, lights, diffusers, grilles, etc., on print of architectural ceiling plan (below the overlay).

41. Color on the overlay, but no indication of any ceiling feature on print of ceiling plan below; indicates a problem which the checker needs to clear up. (Checker provides outlines on the overlay in Colorbrite Medium Red #2126 to indicate problem areas.)

42. Ceiling feature indicated on print of ceiling plan below the overlay, but no color is shown on the overlay; indicates a problem which the checker needs to clear up. (Checker provides outline on the overlay in Colorbrite Medium Red #2126 to indicate problem area.)

43. Checker's notation in Colorbrite Medium Red #2126 applied on the overlay; related entry has been made in the checker's manual.

44. Same as 43, except that yellow added over red indicates that red notation has been copied onto duplicate check-print for communication to others. After problem has been cleared up, checker should erase the yellow/red notation and red outline, and should also correct the ceiling plan overlay for further use.

45. Prismacolor Copenhagen Blue #906 on the overlay, applied over walls and partitions as shown on architectural floor plan.

46. Colorbrite Scarlet #2166 on the overlay, traced over the outline of the structural framing plan of the floor above.

47. Colorbrite Green #2128 on the overlay, traced over plumbing plan; indicates piping at or above ceiling.

48. Eagle Turquoise Yellow on the overlay, applied over the electrical lighting plan; indicates lights at ceiling.

49. Dotted line indicates ducts at ceiling, as shown on HVAC plan under the overlay.

50. Eagle Turquoise Green applied to overlay (over ducts on HVAC plan below) indicates that portion of duct so marked has been reviewed by checker for coordination with architectural (partitions, ceilings, etc.), structural, plumbing, and electrical.

51. Prismacolor Raw Umber #941 applied to overlay (over structural framing as outlined on the overlay per Item 46, above); indicates that portion of structural framing so marked has been reviewed by the checker for coordination with architectural (partitions, ceilings, etc.), HVAC (ducts, etc.), plumbing, and electrical.

52. Colorbrite Medium Red #2126 applied to the overlay; indicates that the checker has made a like-numbered entry (on an 8½ × 11 in. pad) of a comment or a sketch related to the point so marked.

53. Colorbrite Medium Red #2126 applied on the overlay; indicates that the checker has drawn like-numbered schematic sections on an 8½ × 11 in. pad.

54. Described at Item 53, above.

55. Described at Item 53, above.

56. Eagle Turquoise Green applied to the overlay over room names, notes, etc. on HVAC plan below; indicates room names, notes, etc., have been reviewed by the checker.

57. Soft graphite pencil memorandum of beam size (or other desired memoranda); Eagle Turquoise Green is applied over the graphite pencil memorandum when it has been taken into account and is no longer needed. (As an alternative, Colorbrite Green #2128 may be used in lieu of graphite pencil plus light-green tone.)

5 CHECKER'S MANUAL—GENERAL

5-1 CHECKER'S MANUAL

Another basic concept of this system is that the checker, as he proceeds with the checking of check-prints, records all of his comments, sketches, questions, etc., in a standard three-ring, loose-leaf binder. This is called the **checker's manual,** and it is divided into sections (and in some cases into subsections) to receive the information which the checker wishes to record. These sections (and subsections) and how they are used will be described below and in following chapters.

5-2 ADVANTAGES OF THE CHECKER'S MANUAL

This method, in which the checker records comments, sketches, and so on in a checker's manual, has several advantages over the usual procedure, in which the checker marks this information directly on the check-prints. One immediate advantage is that it provides a means whereby all the checker's comments, sketches, and other notes may be consolidated in one place in a logical and accessible arrangement, rather than having them marked all over a multitude of check-prints (often handled in a very haphazard fashion).

A second advantage of recording the checker's information in the checker's manual is that the problem of finding enough unused space on the check-prints for all of the required comments, sketches, and other notes is overcome. It is obvious that many check-prints have little or no unused space which can be utilized for indicating comments, sketches, etc., in the usual manner, but this limitation does not apply at all to the checker's manual; by simply adding more pages to the checker's manual, any number of comments and sketches may be accommodated.

A third advantage of the checker's manual is that the loose-leaf pages may be easily removed from the binder and copied on a standard office copier for transmittal to others (see Chapter 11). In this way the checker can provide others with information without compromising the completeness of his own records (as he retains the originals of all of this information at all times). It is readily apparent that transmitting comments and sketches by copying pages on an office copier is much more efficient than transmitting by forwarding marked check-prints.

A fourth advantage of the checker's manual is that it is a very efficient method of recording checker's questions which are directed to others; the manual yields an orderly arrangement of questions, which greatly facilitates the process of getting them answered. This method of recording the checker's questions in the checker's manual will be described in Chapter 7, and the procedure for getting the questions answered after they have been recorded will be described in Chapter 15.

5-3 SECTIONS IN THE CHECKER'S MANUAL

Plate VII shows all of the sections in the checker's manual and how they should be arranged. Plate VII also indicates additional information (in a tabular form) which aids in visualizing how the sections appear and how they are used. (The abbreviated information in Plate VII will be described in more detail below and in the chapters which follow.)

The first column of Plate VII shows how the **index tabs** for each section in the checker's manual appear. (Each section has a tab-type indexing divider; see Article 5-4.) Each of these index tabs has an abbreviated alphabetical section designation and a distinctive color key, as shown. (The index tab–color key system will be further described in Article 5-5.)

In the second column of Plate VII there is a listing of the title for each of the sections. The third column shows (where applicable) the form of the checker's notation used to relate an item on the check-prints to an entry in the checker's manual. In the fourth column of Plate VII there is a brief description of the type of entry which the checker records in each section (see Chapters 6 and 7).

It may be seen in Plate VII how certain sections of the checker's manual are arranged in groups according to the divisions of the work to which they relate. Thus, comments related to the architectural drawings, sketches related to the architectural drawings, and list of details needed on the architectural drawings, make up the first three sections and are all grouped together. In the same manner, the comments, sketches, and details-needed sections for each of the other divisions of the work (structural, HVAC, plumbing, and electrical) are grouped together. In the same way, the question sections are arranged in groups according to the various divisions of the work to which they relate (plus one miscellaneous group).

This grouping of comments, sketches, details-needed, and question sections according to the divisions of the work to which they relate helps the checker quickly locate any desired section in the checker's manual. In addition, the distinctive color key system applied to the index tabs of the section dividers also helps the checker rapidly locate any desired section.

Following the comments, sketches, and details-needed sections, there are two sections in the checker's manual which do not come under any particular division of the work. These sections are the Checker's Reminders and the Notations sections. (These sections are illustrated in Plate VII and described in Chapter 6.)

The final sections in the checker's manual are the question sections, in which the checker records all of the questions which are directed to various individuals. As mentioned above and shown in Plate VII, the question sections are arranged in groups according to the various divisions of the work to which they relate (plus one group of miscellaneous question subsections). The question sections are divided into subsections; each subsection is assigned to a person to whom questions are directed. These subsections are similar in every respect to the main sections themselves, being separated by the same tab-type indexing dividers and exhibiting the same subsection designations and index tab color keys.

The question subsections are numbered and filed in normal numerical order: AQ1, AQ2, etc. (It should be noted that the question sections shown in Plate VII include only a single, representative subsection in each division of the work; it is likely that in an actual checker's manual there would be a greater number of subsections within each question section.)

5-4 TAB-TYPE INDEXING DIVIDERS

Sections and subsections of the checker's manual should be separated by mylar-reinforced, tab-type **indexing dividers.** The subject of each section and subsection is indicated on the corresponding index tab by an easily read, abbreviated section designation (as shown in Plate VII). In addition, a distinguishing color key is applied to each of the index tabs.

5-5 INDEX TAB–COLOR KEY SYSTEM

To visually differentiate between the various sections of the checker's manual, and also to highlight the grouping of most of the sections according to the division of the work to which they relate, each of the index tabs of the section dividers has a distinctive **color key** applied to it.

As shown in Plate VII, all of the sections related to the architectural division of the work (architectural comments, sketches, details needed, and questions) are highlighted by a light-green color key applied to the index tabs of the section dividers. In like manner, sections related to the structural, HVAC, plumbing, and electrical divisions of the work are highlighted by orange, red, blue, and yellow, respectively.

To help the checker remember which color key represents which division of the work, certain of the colors evoke mental images of the corresponding divisions of the work. Thus, orange (for the structural sections) can remind the checker of orange-painted structural steel, red (for the HVAC sections) can remind him of heating, blue (for the plumbing sections) can remind him of water, and yellow (for the electrical sections) can remind him of an electrical spark.

In addition to the color keys related to various divisions of the work, there are three other color keys in this system: pink, dark green, and violet. The pink color key (for the Notations section) may be thought of as a tone related to the red notations on the check-prints (this helps the checker remember the significance of the pink color key). The remaining color keys are dark green for the Checker's Reminders section and violet for the Miscellaneous Question subsections.

In a short time the significance of each of the color keys becomes second nature to the checker, so that identifying the individual sections or subsections is practically instantaneous.

5-6 TECHNIQUES FOR MAKING ENTRIES IN THE CHECKER'S MANUAL

It is recommended that written entries in the checker's manual be recorded in longhand, using a blue or black ballpoint pen, or another type of marker. (Longhand is recommended because it is much faster than lettering.) Care should be taken to ensure that the written entries are completely legible, not only to the checker who makes them, but also to others who will use the entries.

It is recommended that sketches be made in pencil, because it is erasable and allows portions of the sketch to be erased and corrections made when necessary.

5-7 SIZE, MATERIAL, AND FORMAT OF PAGES IN THE CHECKER'S MANUAL

The pages for both written entries and sketches should be standard 8½ × 11 in. in size, hole-punched for insertion in the checker's manual. Pages for written entries should be wide-ruled, whereas pages for sketches may be ruled, unruled, or cross-section type, according to the checker's preference. For careful sketches which may be traced later, a good grade of bond typewriter paper is recommended; a heavy tracing paper of the proper size might also be used.

The pages in the checker's manual all have a similar format with respect to certain basic information. Figure 5-1 represents a portion of a typical page, showing this basic information. The date shown at the top of the page is the date on which this particular page was first placed in the checker's manual. The project number

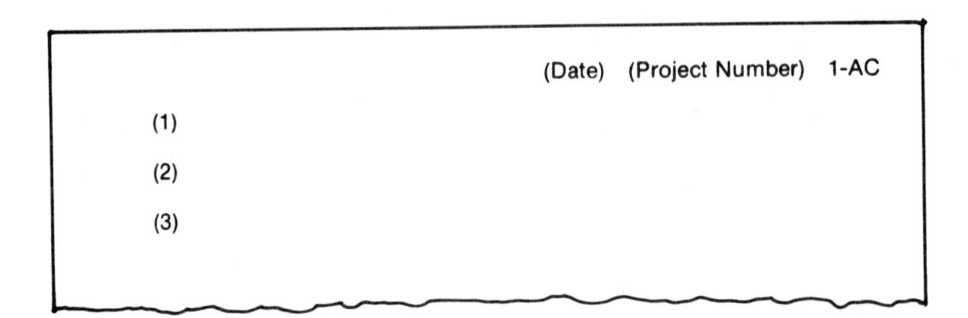

Figure 5-1

is also shown at the top of the page. The number 1 in the upper-right corner of the page is its chronological number, indicating that it is page 1 of this section. Subsequent pages 2, 3, 4, etc., are placed over this page 1 in reverse chronological order, with the latest page on top. This means that the latest page is always the one immediately behind the indexing divider.

The AC in the upper-right corner of the page is the **section designation,** indicating the section of the checker's manual in which the page appears. AC indicates the Architectural Comments section; other section designations are AS, AD, SC, SS, SD, etc., indicating the sections shown in Plate VII. (In addition to appearing in the upper-right corner of each page, the appropriate section designation appears on the index tab of the section divider, as shown in Plate VII.)

The indications (1), (2), and (3) are the chronological numbers of the entries which the checker records. These entry numbers start with (1) on the first page of each section and run consecutively through all of the succeeding pages of that section. (The entries themselves are not shown in Figure 5-1, but will be described in detail below.)

5-8 DRAWING LOCATION KEYS

With the exception of entries in the comments sections, where an entry in the checker's manual relates to a particular drawing or to a specific location on a particular drawing, the checker provides a **drawing location key** which serves to key that entry both to the proper drawing and to a specific location on the drawing when applicable.

The drawing location keys in each case include the number of the drawing to which the entry refers. In addition, if an entry relates to a specific location on a drawing, additional location information is provided by the checker. This additional location information could take a number of forms, according to the situation. One form might be a reference to a specific item on the drawing, such as "front elevation," "Space 145," "near column 10," or "near door 76." Another form might be a reference to a particular section or detail number. Still another form might be a reference to a particular quadrant of the drawing, with U/L indicating the upper-left quadrant, U/R indicating the upper-right quadrant, etc. For increased precision, the checker might wish to include more than one of these forms of location information.

When a drawing location key is needed, it should be placed in parentheses immediately following the entry number. (Entry numbers are described in Article 5-7.) The following example shows an entry number with its related drawing location key:

(12) (A-4, Detail 7, U/L)

This indicates that entry number (12) relates to drawing number A-4, and specifically to Detail 7, which appears in the upper-left quadrant of the drawing. Drawing location keys are used by the checker and others in a number of other ways which will be described in succeeding chapters.

5-9 PROCEDURES FOR RECORDING AND USING ENTRIES IN THE CHECKER'S MANUAL

Procedures for recording and using entries in the checker's manual will be described in the chapters which follow.

6 CHECKER'S MANUAL—COMMENTS, SKETCHES, DETAILS NEEDED, AND OTHER SECTIONS

6-1 COMMENTS, SKETCHES, DETAILS NEEDED, AND OTHER SECTIONS

The Comments, Sketches, Details-Needed, Checker's Reminders, and Notations sections in the checker's manual were described in general in Chapter 5. This chapter describes in detail the procedures used to make entries in each of those sections. (How the entries in various sections are used will be described in subsequent chapters.)

6-2 ARCHITECTURAL COMMENTS

Let us assume the checker is checking an architectural master check-print and he observes something about which he wishes to make a comment. As this is an architectural drawing, the checker needs to communicate his comment to the architectural personnel. To do this, the checker records an entry in the Architectural Comments (AC) section of the checker's manual for later transmittal to the architectural personnel.

To record this entry, the checker first turns in the checker's manual to the indexing divider having the section designation AC on the index tab. Immediately behind this indexing divider, the checker finds the latest page in the section (see Article 5-7).

Let us assume that the latest page in the section appears as in Figure 6-1, with the date; project number; page number and section designation (1-AC); and the entry numbers (1), (2), and (3) (see Article 5-7).

(Date) (Project Number) 1-AC

(1) Add floor-to-floor dimensions on all elevations; see sections for these.

(2) Shown as double doors, but plan shows single door. Which is correct?

(3) Add note: "See HVAC plan for sizes of all door grilles."

Figure 6-1

Remark: It should be noted that no drawing location keys are included with these entries in the Architectural Comments section. The reason for this is that the checker and others use the notations on the check-prints to direct them to the entries, rather than the other way around. The checker may, however, wish to include drawing location keys as a double-check (in case he inadvertently fails to place a notation on the check-prints, for instance). If the checker does elect to include drawing location keys with entries in the comments sections, they would appear as described in Article 5-8. (Drawing location keys are included in all of the other sections of the checker's manual.)

Having found the page 1-AC in the Architectural Comments section, the checker now proceeds to record whatever comment he wishes to make regarding the relevant architectural master check-print. Let us assume that he is checking architectural master check-print A-9, and he observes that a wall section shows a brick veneer as the facing for a concrete wall, but there is no indication of any method of securing the brick veneer to the concrete wall. To record his comment regarding a method of securing the brick veneer to the concrete wall, the checker adds a new entry on page 1-AC, below the previous entry. This new entry is number (4), and it appears as follows:

(4) Provide vertical dovetail slots, 16 in. o.c., in face of concrete wall; provide corrugated dovetail wall ties 16 in. o.c. each way to secure brick veneer

Either before or after recording this new entry number (4) in the checker's manual, the checker places the notation AC4 in red pencil on check-print A-9, in the area of the omitted dovetail slots and wall ties (thus referencing that point on the drawing to the new entry).

Remark: As described above, if the checker wishes, he may include a drawing location key with the entry, as a possible double-check later. In this example, the drawing location key is (A-9).

6-3 ARCHITECTURAL SKETCHES

Let us assume that the checker is checking an architectural master check-print and that he observes something about which he wishes to make an explanatory sketch. As this is an architectural drawing, the checker needs to communicate his sketch to the architectural personnel. To do this, the checker records an entry (a sketch) in the Architectural Sketches (AS) section of the checker's manual, for later transmittal to the architectural personnel.

To record this entry, the checker first turns in the checker's manual to the indexing divider with the section designation AS on the index tab. Immediately behind this indexing divider, the checker finds the latest page in the section.

Let us assume that the latest page in the section appears as shown in Figure 6-2, with the date; project number; page number and section designation (1-AS); and entry numbers (1), (2), and (3). The indications (A-2, Column 5) and (A-3) are drawing location keys which are used to key sketches to the drawings.

Following the sketch number (and the drawing location key, when it occurs), the checker should provide a descriptive title for the sketch. Where a sketch is drawn to scale, the scale should be indicated with the title.

It will be noted that the sketch number and the title of each sketch is shown above the sketch; the reason for this is that the checker, starting at the top of the page, can go ahead and indicate the title of the sketch first, before he knows exactly how big the sketch will be. This is, of course, the checker's option. As an alternative, the checker may prefer to start the sketches at the bottom of the page,

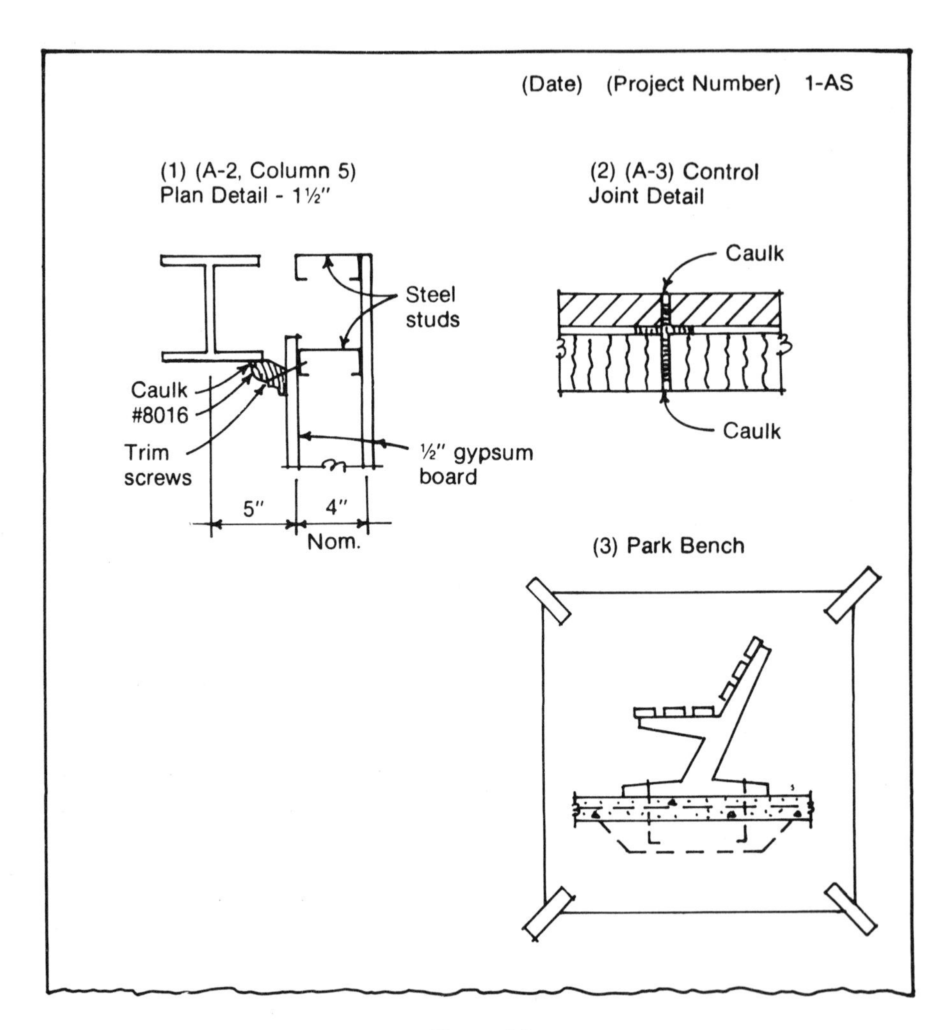

Figure 6-2

with the sketch number and the title of each sketch below. Again, this is the checker's option.

Sketch number (3) is shown to illustrate a technique in which the checker does not actually make the sketch but instead uses a print of a detail which he has cut from a previous project and taped onto the page of sketches. In a variation of this, the checker might cut the detail from a file of standard details, or he might use a detail cut from a manufacturer's printed data sheet.

Having found the page 1-AS in the Architectural Sketches section, the checker now proceeds to record whatever sketch he wishes to make regarding the relevant architectural master check-print.

Let us assume that the checker is checking architectural master check-print A-4, and he observes that some critical dimensions have been omitted from a window sill detail. To record a sketch indicating the desired dimensions of the window sill, the checker adds a new sketch on page 1-AS, near the previous sketch. This new sketch is number (4), its drawing location key is (A-4), and it shows all of the dimensions which the checker wants to add to the window sill detail.

Either before or after making the new sketch number (4) in the checker's manual, the checker places the notation AS4 in red pencil on check-print number A-4, in the area of the window sill detail in question, thus referencing that point on the drawing to the new entry.

Remark: The extent of the checker's sketches, how complete they are, whether they are drawn freehand or not, whether they are drawn to scale or not, etc., depends on how much time the checker has and on other factors. Unless he has a very generous amount of time, the checker should not be expected to work out every problem in complete detail. If time is limited, or if other personnel might be more conversant with certain problems, the checker should use the comment or the question sections of the checker's manual to pass those problems along to the proper persons.

As described in Article 5-7, the sketches in the checker's manual can be made on ruled, unruled, or cross-section paper, according to the checker's preference. For careful sketches which may be traced later, a good grade of bond typewriter paper is recommended, or a heavy tracing paper might be used.

If the checker has occasion to make any drawings larger than an 8½ × 11 in. page size, they should be made on tracing paper so that they can be printed for the checker's files and for transmittal of duplicate prints to others. The checker may wish to key these larger drawings to checker's notations on the check-prints by tying the drawing numbers into the numbering system of the sketches in the checker's manual.

Remark: It is suggested that the first page in each sketch section (and supplemental pages as required) be utilized to indicate rough **key plans,** and that each subsequent sketch in that section be flagged on the key plans by a section-cut (or other indication) and a reference to the sketch number.

6-4 ARCHITECTURAL DETAILS NEEDED

Let us assume the checker is checking an architectural master check-print and he observes an item which needs further detailing. As this is an architectural drawing, the checker wants to communicate the need for this further detailing to the architectural personnel. To do this, the checker records an entry in the Architectural Details-Needed (AD) section of the checker's manual, listing the item and describing the added detailing which is needed. This entry is later transmitted to the architectural personnel.

To record this entry, the checker first turns in the checker's manual to the indexing divider with the section designation AD on the index tab. Immediately behind this indexing divider, the checker finds the latest page in the section.

Let us assume that this latest page in the section appears as in Figure 6-3, with the date; project number; page number and section designation (1-AD); and the entry numbers (1), (2), and (3). The indications (A-7) and (A-5) are drawing location keys which are used to key entries to the drawings.

Having found the page 1-AD in the Architectural Details-Needed section, the checker proceeds to record an entry listing the added

(Date) (Project Number) 1-AD

(1) (A-7) Canopy at entrance on North Elevation needs to be detailed.

(2) (A-5) Show all window details on Sheet A-5.

(3) Need details of all doors.

Figure 6-3

detailing needed on the item on the architectural master check-print which he is checking. Let us assume that the checker is checking architectural master check-print A-8, and he observes that a cabinet shown in Space 416 apparently has not been detailed. To record his entry listing this cabinet as in need of detailing, the checker adds a new entry on page 1-AD, below the previous entry. This new entry is number (4), its drawing location key is (A-8, Space 416), and it appears as follows:

(4) (A-8, Space 416) Cabinet: Need elevation and sections

Either before or after recording the new entry number (4) in the checker's manual, the checker should place the notation AD4 in red pencil on check-print A-8, on (or near) the cabinet which needs detailing, thus referencing that point on the drawing to the new entry. (If an entry does not refer to a specific point on a drawing, this step would be omitted.)

6-5 ENGINEERING COMMENTS, SKETCHES, AND DETAILS NEEDED

If the checker is checking a structural, HVAC, plumbing, or electrical master check-print and wants to make any comments or sketches, or if he wishes to list additional details needed in relation to that drawing, such comments, sketches, and listing of details needed would be recorded in the appropriate structural, HVAC, plumbing, or electrical sections of the checker's manual.

The procedures used in recording comments, sketches, and lists of details needed in the engineering sections of the checker's manual are similar to the procedures used in the architectural sections. The procedures described in the preceding articles may be used verbatim for recording comments, sketches, and lists of details needed in the engineering sections, except that wherever the word "architectural" appears, the appropriate word "structural," "HVAC," "plumbing," or "electrical" is substituted. Also, wherever the abbreviation "A" appears, the appropriate abbreviation "S," "H," "P," or "E" is substituted.

6-6 CHECKER'S REMINDERS

Let us assume the checker is checking *any* master check-print—regardless of whether it is an architectural, structural, HVAC,

(Date) (Project Number) 1-CR

(1) (A-5, Detail 4) Entrance — hold checking this until later, when I have print of Sheet A-9

(2) (A-1) Need to get print of topographical survey and check all elevations on Site Plan against survey.

(3) Make general check of building code before plans are issued.

Figure 6-4

plumbing, or electrical check-print—and he wishes to record a miscellaneous reminder directed to himself. To do this, the checker records an entry in the Checker's Reminders (CR) section of the checker's manual.

To record this entry, the checker first turns in the checker's manual to the indexing divider having the section designation CR on the index tab. Immediately behind this indexing divider, the checker finds the latest page in the section.

Let us assume that this latest page in the section appears as shown in Figure 6-4, with the date; project number; page number and section designation (1-CR); and the entry numbers (1), (2), and (3). The indications (A-5, Detail 4) and (A-1) are drawing location keys which are used to key entries to the drawings.

After finding the page 1-CR in the Checker's Reminders section, the checker proceeds to enter whatever reminder to himself he wishes to record. Let us assume that the checker is checking structural master check-print S-3, and that he checks a detail showing a shelf angle carrying some limestone veneer. The checker is concerned about the particular method used to secure the limestone on the shelf angle, but he wishes to check some data on limestone anchorage before he discusses the problem with the structural engineer.

To record a reminder to himself to check this limestone data, the checker adds a new entry on page 1-CR, below the previous entry. This new entry is number (4), its drawing location key is (S-3), and it appears as follows:

(4) (S-3) Check limestone anchorage against limestone handbook

Either before or after recording the new entry number (4) in the checker's manual, the checker should place the notation CR4 in red pencil on check-print S-3, in the area of the limestone detail which he is questioning, thus referencing that point on the drawing to the new entry. (If an entry does not refer to a specific point on a drawing, this step would be omitted.)

Remark: In the above example, the checker's reminder referred to an item on a check-print. It should be understood that checker's reminders are not limited to references to the drawings, but may refer to any subject about which the checker wishes to be reminded.

6-7 NOTATIONS

Let us assume that the checker is checking *any* master check-print and that he finds that the check-print is too crowded for all of the checker's notations which he wishes to indicate regarding a certain item. To overcome this difficulty, the checker places a *single* checker's notation (as described in Article 4-5 and also below) on the check-print. This single notation serves to refer to an entry in the Notations (NO) section of the checker's manual. At this entry, the checker records as many notations as he wishes regarding the item in question.

To record this entry, the checker first turns in the checker's manual to the indexing divider with the section designation NO on the index tab. Immediately behind this indexing divider, the checker would find the latest page in the section.

Let us assume that this latest page in the section appears as in Figure 6-5, with the date; project number; page number and section designation (1-NO); and the entry numbers (1), (2) and (3). The indications (A-2) and (A-5) are drawing location keys which are used to key entries to the drawings. (The (") indicates that the previous drawing location key is repeated.)

The notations P3)E)R —\/, P1 — S)G)CR20, etc., are the checker's notations which were entered in the checker's manual because there was not enough space on the master check-print. (The mean-

(Date) (Project Number) 1-NO

(1) (A-2) P3)E)R —\/

(2) (A-5) P1 – S)G)CR20

(3) (") —\|/ P4)AC19

Figure 6-5

ings of these notations are described in Chapter 4.) These checker's notations are used in the same way as if they had been placed directly on, or near, an item on a master check-print.

After finding the page 1-NO in the Notations section, the checker proceeds to record whatever checker's notations he wishes to make regarding the item in question on the master check-print.

Let us assume that the checker is checking architectural master check-print A-7, and he observes a vertical exhaust duct on the second-floor plan that needs to be checked on the following plans: first floor, third floor, first-floor ceiling, second-floor ceiling, third-floor ceiling, roof, structural, HVAC, plumbing, and electrical. Let us also assume that the checker has made an architectural comment number (14) and an architectural sketch number (3) regarding this duct. As we have assumed that this master check-print A-7 is so crowded that the checker cannot find room for all of the checker's notations which he wishes to apply, he must record them in the checker's manual.

The checker adds a new entry on page 1-NO, below the previous entry; this new entry is number (4), its drawing location key is (A-7), and it appears as follows:

(4) (A-7) P1)P3)C1)C2)C3)R —\ | / AC14)AS3

Either before or after recording the new entry number (4) in the checker's manual, the checker should place the notation NO4 in red pencil on check-print A-7, in the area of the vertical duct in question, thus referencing that point on the drawing to the new entry.

7 CHECKER'S MANUAL—QUESTION SECTIONS

7-1 QUESTION SECTIONS

Question sections in the checker's manual are those sections in which the checker records questions to be directed to various persons. Question sections were described in general in Chapter 5, but this chapter will describe in detail the procedures used in making entries in the question sections.

7-2 GENERAL

The checker's manual is provided with six different question sections, entitled: Architectural Questions, Structural Questions, HVAC Questions, Plumbing Questions, Electrical Questions, and Miscellaneous Questions. These question sections are arranged in the checker's manual as shown in Plate VII.

All of the checker's questions directed to persons involved in the architectural division of the work are entered in the Architectural Questions section; all questions directed to persons involved in the structural division of the work are entered in the Structural Questions section; and so on for the questions directed to the HVAC, plumbing, and electrical divisions of the work. In addition, where the checker's questions are directed to persons not directly involved in one of the architectural or engineering divisions of the work—such as owner's representatives, contractor's representatives, building officials, or special consultants—such questions are recorded in the Miscellaneous Questions section.

Each of the six question sections are subdivided into a number of individual subsections according to the number of persons to whom the checker has addressed questions. For example, if the checker has directed questions to four different persons involved in the architectural division of the work, he would provide four corresponding individual subsections in the Architectural Questions section, and each of these subsections would be used to record the relevant checker's questions. All of the other question sections (structural, HVAC, etc.) would be similarly subdivided into individual subsections according to the number of persons involved in each case.

7-3 QUESTION SUBSECTION INDEXES

When the checker wants to record a question directed to a particular person, he must be able to locate rapidly the specific subsection in which questions directed to that person are to be recorded. To easily locate a specific subsection, the checker provides each of the six question sections with a **question subsection index.**

The six question subsection indexes are called the Architectural Question subsections, Structural Question subsections, HVAC Question subsections, Plumbing Question subsections, Electrical Question subsections, and the Miscellaneous Question subsections.

To illustrate how the various question subsection indexes work, an example of a typical Architectural Question subsection index is shown in Figure 7-1, with the date, project number, and page number and section designation (1-AQS) (see Article 5-7).

The indications AQ1, 2, 3, and 4 are the subsection designations of the Architectural Question subsections which follow this index. These subsection designations are indicated on the index tabs of the corresponding question subsections which follow the index, shown as AQ1, AQ2, etc. (see also Plate VII). The subsection designations also appear on each page of entries in corresponding subsections, as described in Article 7-8.

As in Figure 7-1, each subsection designation is followed by the name of a person. All of the checker's questions directed to a particular person are recorded in the subsection which corresponds to that person's name in the index. For example (using the

(Date) (Project Number) 1-AQS

Architectural Question Subsections

AQ1—Mary Smith (Project Architect)

2—Joe Jones (Project Manager)

3—Bill Williams (Job Captain)

4—Jim Baker (Job Captain)

Figure 7-1

subsection index shown in Figure 7-1), all of the checker's questions directed to Mary Smith are recorded in the question subsection designated AQ1, and so on for the other persons named.

Mary Smith, Joe Jones, etc., are the names of those persons involved in the architectural division of the work to whom the checker directs questions. The order of this list is completely arbitrary; other names may be added at any time if the checker directs questions to additional persons in the architectural division of the work.

The titles project architect, project manager, etc., are informal job descriptions of the various persons listed; these are noted by the checker solely for his own benefit, to aid in identifying each person.

7-4 ENGINEERING AND MISCELLANEOUS QUESTION SUBSECTION INDEXES

Indexes for the Engineering and the Miscellaneous Question subsections follow the same format as the Architectural Question subsection index shown in Figure 7-1, except that where the word "architectural" appears, the appropriate word—"structural," "HVAC," "plumbing," "electrical," or "miscellaneous"—is substituted. Also, wherever the abbreviation "A" appears, the appropriate abbreviation—"S," "H," "P," "E," or "M"—is substituted.

7-5 QUESTION SUBSECTION INDEXES INVOLVING LARGE ORGANIZATIONS

Figure 7-1 shows an index consisting of a simple listing of persons, along with the question subsection designations. If the checker has

(Date) (Project Number) 1-MQS

Miscellaneous Question Subsections

MQ1—Jack Ramsey Corporation (Owner)

MQ1-1—Jack Ramsey (President)
2—Sam Roberts (Project Manager)
3—Will Bush (Assistant Project Manager)

MQ2—Lloyd Construction Company (Contractor)

MQ2-1—Martin Lloyd (President)
2—Paul Johnson (Project Manager)
3—Doug Wade (Estimator)

MQ3—City of (name of city) Administration

A—Planning and Zoning Department

MQ3-A-1—Don Campbell (Head of Department)
2—Doyle Lamar (Assistant)

B—Building Department

MQ3-B-1—Charles Boyd (Permits)
2—Grady Cooper (Fire Marshal)
3—Brad Arnold (Inspections)

C—Traffic Engineering

MQ3-C-1—Dan Hawkins (Head of Department)
2—Terry Dillard (Assistant)

MQ4—Ralph Johnson (Acoustical Consultant)

Figure 7-2

occasion to direct questions to persons in large organizations, it is helpful to make up an index which provides an organizational grouping of persons, rather than merely listing the persons individually. Figure 7-2 represents a hypothetical Miscellaneous Question subsection index, illustrating the form which should be used when large organizations are involved.

In this hypothetical index, each organization is identified by a basic designation (MQ1, MQ2, MQ3, etc.). If an organization were further divided into departments, each of those departments would be identified by a letter (A, B, C, etc.). And finally, each subsection is given a number and is assigned to a person in the organization. Thus, by using the basic designation of the organization, the letter indicating the department (where necessary), and the individual subsection number, each person in an organization may be related to a specific subsection designation. This subsection designation then serves to indicate the proper subsection in which to record questions directed to that person. In the same manner as described in Article 7-3, the subsection designations in an index involving large organizations also appear on each page of entries in corresponding subsections following the index, as well as on the index tabs of those subsections.

In a question subsection index involving large organizations, the checker should skip several lines between different organizations, and between different departments in the same organization, to provide space for the addition of other names if it becomes necessary later.

Question subsection indexes involving large organizations may, of course, also include individuals not associated with a large organization. An example of this is shown in Figure 7-2 at the designation "MQ4—Ralph Johnson (Acoustical Consultant)."

7-6 PLACEMENT OF QUESTION SUBSECTION INDEXES IN THE CHECKER'S MANUAL

The various question subsection indexes are placed in the checker's manual immediately behind the proper subsection indexing divider in each case. (The order of placement is shown in Plate VII.)

7-7 INDEX TABS FOR QUESTION SUBSECTIONS

The index tabs for the indexing dividers for the question subsections have subsection designations and index tab color keys (see Chapter 5 and Plate VII). Immediately behind each index are placed the subsections to which the index relates. The question subsections are numbered and filed in normal numerical order; for example, AQ1, AQ2, etc.

7-8 CHECKER'S QUESTIONS

To illustrate how the question sections of the checker's manual are used, let us assume the checker is checking *any* master check-

(Date) (Project Number) Bill Williams 1-AQ3

(1) (A-3, Space 207) Range, is this gas or electric? It is not shown on plumbing or electrical.

(2) (A-9) Do you want to show entrance details on Sheet A-9?

(3) When are we going to meet with structural engineer?

Figure 7-3

print—regardless of whether it is an architectural, structural, HVAC, plumbing, or electrical check-print—and that the checker wishes to direct a question about that check-print to Bill Williams, a job captain in the architectural division of the work. (Actually, checker's questions are not limited to the drawings, but may be concerned with any subject; a question regarding a drawing is used here merely as an example.) Since Bill Williams is in the architectural division of the work, a question to him is recorded in the Architectural Questions section of the checker's manual.

To record this question, the checker first turns in the checker's manual to the indexing divider having the designation AQS on the index tab (indicating the Architectural Question subsection index). Immediately behind this indexing divider, the checker finds the subsection index; let us assume that the index is as shown in Figure 7-1. The checker then locates the name Bill Williams, and notes that Williams has been assigned to subsection AQ3.

To find subsection AQ3, the checker examines the index tabs of the various Architectural Question subsections (which should be arranged in normal numerical order), and when he comes to the divider having the index tab AQ3, he finds the latest page in subsection AQ3 immediately below the divider (see Article 5-7).

Let us assume that the latest page in subsection AQ3 appears as shown in Figure 7-3, with the date; project number; page number and subsection designation (1-AQ3); and the entry numbers (1), (2), and (3) (see Article 5-7). Also, the name "Bill Williams" is noted at the top of the page. The indications (A-3, Space 207) and (A-9) are drawing location keys which are used to key entries to the drawings (see Article 5-8).

After finding this page 1-AQ3 in subsection AQ3 of the Architectural Questions section, the checker proceeds to record whatever questions he wishes to direct to Bill Williams. As shown in Figure 7-3, the checker would skip one or two lines between each of the questions. This is done in order to provide space to record answers to the questions at a later time, as will be described in Chapter 15.

Let us assume that the checker is checking architectural master check-print A-12, and he notices that detail number 7 shows a balcony rail which is 36 in. in height; the checker believes the building code may require a rail 42 in. high, and he wishes to ask Bill Williams if this has been checked in the building code.

To record this question for Bill Williams, the checker adds a new entry on page 1-AQ3 below the previous entry. This new entry is number (4), its drawing location key is (A-12, Detail 7), and it appears as follows:

(4) (A-12, Detail 7): Balcony rail; shown 36 in. high. Should this be 42 in. high? Have you checked this in the building code?

The checker does *not* place any notation in red pencil on check-print A-12, near detail 7, to reference that point on the drawing to the new entry. This is different from the procedure regarding entries in all other sections of the checker's manual. The reason for this is that the checker uses the entries in the question sections in a somewhat different manner from the entries in other sections. How these entries in the question sections are used will be described in Chapter 15.

7-9 ENGINEERING AND MISCELLANEOUS QUESTION SECTIONS

The procedures used to record questions in the Engineering and the Miscellaneous Question sections of the checker's manual are similar to those used to record questions in the Architectural Question section. Thus, the procedures described in the preceding article may be used verbatim for recording questions in the Engineering Question sections, except that wherever the word "architectural" appears, the appropriate word—"structural," "HVAC," "plumbing," "electrical," or "miscellaneous"—is substituted. Also, wherever the abbreviation "A" appears, the appropriate abbreviation—"S," "H," "P," "E," or "M"—is substituted.

7-10 QUESTION SECTIONS IN SEPARATE BINDER

If a project is of any appreciable size, the question sections should be placed in a separate three-ring binder. The reason for this is to facilitate the use of the question sections at checker's conferences (see Chapter 15). At checker's conferences, only the question sections are needed, so that having them in a separate binder is the most efficient arrangement.

8 CHECKER'S MANUAL—HOW TO USE SECTIONS

8-1 HOW TO USE SECTIONS IN THE CHECKER'S MANUAL

After the checker makes entries in the various sections of the checker's manual, he uses the entries in a variety of ways. The following is a summary of how the entries are used.

Copies of the checker's comments and sketches regarding the drawings, and the list of details needed, are transmitted (along with duplicate check-prints) to personnel in the various divisions of the work for use in revising the drawings (this transmittal will be described in Chapter 11).

The Checker's Reminders section is used by the checker as a checklist of reminders of actions which must be taken regarding items on the drawings (or elsewhere). Where an entry refers to an item on a drawing, the checker is able to find that item by using the drawing location key included with such entries (see Article 5-8).

The Notations section is used by the checker to record checker's notations which he cannot place on the master check-prints due to space limitations. The notations in this section are used in exactly the same way as the checker's notations which are placed directly on the master check-prints.

The question sections are used to get the checker's questions answered by personnel in various divisions of the work, and also by any other persons to whom the checker has addressed questions. (The procedure used in getting questions answered will be described in Chapter 15.)

8-2 VERIFYING THAT NECESSARY ACTIONS HAVE BEEN TAKEN

After the pages of entries in the various sections of the checker's manual have been used to get revisions made, questions answered, etc., the checker should review the pages of entries and verify that all necessary actions have been taken in accord with each of the entries.

The checker carefully examines every aspect of each entry to see that it has been fully taken into account. As an aid in doing this, the checker should apply light-green pencil over all written entries as he reviews them, and over all sketches and notes (in the same manner as the checker's color code is applied to the check-prints), to indicate that all items have been taken into account and require no further attention.

When the checker has verified that an entry has been taken into account and requires no further action, he should draw a vertical green line through that entry, as illustrated in Plate III. (Colorbrite Green #2128 is an appropriate green for this purpose.) This vertical green line should be located at approximately the center of the page for each entry, so that the green line drawn through one entry will connect to the green line drawn through the entries above and below. In this way, when all entries on a page have been taken into account and have had vertical green lines drawn through them, the connected segments of green line will run unbroken from top to bottom of the page. This unbroken green line then provides an instantly recognizable signal that the entire page has been taken into account and requires no further attention.

8-3 INACTIVE CHECKER'S MANUAL

On large projects the checker provides himself with an additional three-ring loose-leaf binder which is called the **inactive checker's manual**. This binder is used to file all inactive pages from the checker's manual. When all the entries on a page in the checker's manual have been taken into account, and a vertical green line has been applied through every entry, that page should be retired and placed in the inactive checker's manual.

The purpose of having this inactive checker's manual is to enable the checker to keep the active checker's manual free of inactive material, which can seriously impede efficiency if it becomes at all voluminous. At the same time, the inactive checker's manual allows the checker to retain the inactive material in case it is needed later. All pages in the inactive checker's manual should be kept in careful order, by using the same arrangement as with the checker's manual.

8-4 FINAL CHECK OF PAGES IN CHECKER'S MANUAL

Before final release of drawings, the checker should go through all pages of the checker's manual and be sure that all entries there have been taken into account.

9 SINGLE-SHEET CHECK

9-1 SINGLE-SHEET CHECK

A basic concept of this system of checking is a technique whereby the checker can check a single sheet of drawings without referring to any other sheet until a later time. This **single-sheet check** technique allows the checker to proceed with the checking even though only one sheet is available to check.

Generally, any sheet has both items which *can* be checked independently of any other sheet and items which *cannot* be checked without referring to another sheet. This single-sheet check provides a technique for handling both types of items.

9-2 PROCEDURE

To use this technique, the checker selects any desired item on a check-print and subjects it to an intensive review, carefully analyzing the item from several aspects. First, the checker considers the general appropriateness of the item as it is shown, and looks for obvious mistakes or incongruities. Second, he considers the relationship of the item to all other parts of the project; for example, would the item affect any elements of the architectural work and would it have any structural, HVAC, plumbing, or electrical work related to it? From this, the checker can determine what other points on the drawings need to be checked regarding the item. And last, the checker considers whether there are any comments or questions he would like to direct to others regarding the item, or whether there are any sketches he wants to make regarding the item.

As an example of the analysis of an item, let us assume that the checker observes an item on an architectural second-floor plan which is noted as "Range, hood over." First, the checker examines the indication of the item for any obvious mistakes or incongruities. Does it appear to be properly located? Does the size appear to make sense? Next, the checker considers what divisions of the work would be affected by the item; from this he would determine what other sheets of drawings need to be checked regarding the item. In this example, the checker checks the following: structural plan of floor system for support of the range; architectural and structural drawings for duct and piping openings through floors, roof, or walls; HVAC, for hood details and ducts; plumbing, for any gas piping; and electrical, for any electrical connections and lighting. In addition, the checker needs to check the second-floor ceiling plan for the hood and any blow-ups (larger-scale drawings) of the area.

As the checker proceeds with the analysis of an item, he makes a graphic record to indicate each mental observation he makes regarding the item. He produces this graphic record by means of checker's notations in red pencil on (or near) the item on the check-print, and by written entries (related to the notations) which he makes in the checker's manual. (These notations and the related entries are made as described in Chapters 4 through 7.)

In this example of "Range, hood over," the checker applies notations indicating checking of structural, HVAC, plumbing, and electrical plans, as well as of the second-floor ceiling plan and any blow-up of the area. In addition, let us suppose for the purpose of illustration that the checker has made sketch number 19 in the Architectural Sketches section of the checker's manual. In this example, all of the notations applied to the check-print might appear thus:

_\I/C2)B)AS19

If, as in the example above of the "Range, hood over," the checker's analysis of the item indicates that other related points on the drawings need checking, or the checker has comments, the checker would first make all notations in red as described above (and enter desired comments and sketches in the checker's manual). Then he would apply green on the check-print over the item (or around its perimeter if the item is large) *not to indicate that the item is necessarily correct as shown*, but to indicate that the item has been reviewed and is *provisionally accepted, subject to any changes arising from the checker's follow-up on the actions indicated by the notations*. (This procedure of applying green to provisionally accepted items is described in Chapter 3.)

If the checker's analysis of an item indicates that the item is apparently correct as it is shown and needs no further checking at other points on the drawings, he would apply green on the check-print over the item (or around its perimeter if the item is large); this indicates that the item is *tentatively accepted* as being correct as shown and that it requires no further attention, unless other information regarding the item is subsequently brought to light by the checking of other drawings.

The above process, of analyzing an item, applying red notations, entering comments in the checker's manual, and finally applying green over the item (or around its outline) is repeated for each item on the sheet until all items have had green applied to them.

9-3 EXCEPTIONS

In the single-sheet check technique, there are certain items which should be left for later checking, using the multiple-sheet cross-check technique, which will be described in Chapter 10.

Examples of these exceptions are door details, millwork details, and many other details. Door details need to be checked against door schedules, floor plans, and finish schedules; millwork details need to be checked against floor plans, sections, and finish schedules. In these cases it is better not to take the time to check all these items and make the notations in red, a requirement of the single-sheet check technique, but to wait and check them using the multiple-sheet cross-check technique.

Other items which are not checked by this single-sheet check technique are items in the ceiling, whether they are shown on the architectural ceiling plans, on the HVAC plans (diffusers, grilles, etc.), or on the electrical plans (lights). All ceiling items should be checked by use of the ceiling plan overlay technique described in Chapter 18.

In addition, such items as windows, doors, and other features shown on the elevations and floor plans should not be checked by the single-sheet check technique, but should be checked against each other (on the elevations and floor plans) by use of the two-sheet cross-check technique described in Chapter 10.

9-4 CHECKING INCOMPLETE CHECK SETS

The single-sheet check technique eliminates a major problem encountered when using the usual method of checking, in which the checker selects an item on one sheet and then hunts through the entire check-set for all other sheets on which that item should appear, in order to check the item on each of the other involved sheets. The problem here is that the check-set must be complete for the usual method to work—if sheets are missing, it is impossible to completely check an item.

By using the single-sheet check technique, however, the checker may check each sheet independently without referring to any other sheet until a later time. Thus, sheets missing from the check-set have no effect on the checker's ability to proceed with the checking of any available sheets. When the previously missing sheets become available, they may be checked by the single-sheet check technique; then all available sheets may be cross-checked by the methods described in Chapter 10.

9-5 GRAPHIC CHECKLIST

Another advantage of the single-sheet check technique is that the notations in red regarding each item form a miniature graphic checklist for the checker, a feature which can be very helpful. In the first place, the checker is aided in making a mental analysis of each item by the fact that he records (by means of the notations) each decision about what further actions are required regarding that item. This allows the checker to proceed very carefully in his analysis, since the result of each decision may be studied and reviewed as often as desired. Having this graphic record relieves the checker from relying on his memory; he is able to glance at an item at any time and see exactly what required actions he has indicated regarding the item, and if he feels that further actions are required, further notations may be added at any time.

In addition, as soon as the implied directive of a notation has been carried out, the checker applies green pencil over that notation to indicate that it requires no further attention. Again, this frees the checker of having to rely on his memory. He does not have to ask himself, "Did I check this item on the HVAC drawings?" or "Did I check this item on the structural drawings?" By using this graphic system, the checker can tell at a glance what required actions were noted regarding each item and the exact status of those actions. If a red notation has had green applied over it, the indicated action has been taken; if a red notation remains red, it still needs attention.

9-6 SPACE-BY-SPACE CHECK

As an alternative to **item-by-item checking** and the application of checker's notations as described above, the checker might use a somewhat less precise approach, in which he considers an entire space at one time instead of checking each item separately at the single-sheet check stage.

For example, let us assume that the checker observes a restroom on an architectural floor plan. Instead of applying a checker's notation on each individual fixture to indicate that the fixture needs to be cross-checked against the plumbing drawings, the checker could apply a single notation (in the immediate area of the space name) to indicate that all the fixtures in that space need to be checked against the plumbing drawings. If other items in that space need to be checked against the electrical or other engineering drawings, the checker applies the appropriate notations, again in the area of the space name. When checking of all items called for by the single notation in a space has been accomplished, the checker applies a light-green tone over that notation.

While this **space-by-space check** saves time, it is obvious that extra care has to be taken to ensure that each item is cross-checked. With the item-by-item check, including separate notations for each item, there is less chance that an item would fail to be cross-checked at all the necessary points on other drawings.

9-7 ELEVATION-BY-ELEVATION CHECK

In the same manner as the space-by-space check described above, the checker could use a single notation to indicate that all items on the plan which appear on the exterior (on one side of the building) should be checked against the elevation drawings of that side of the building. This is known as an **elevation-by-elevation check**.

To do this, the checker places the checker's notation E on the plan, adjacent to the exterior wall in question, and perhaps includes an arrow from the notation pointing toward the side of the building. When all items on the plan have been checked against the appropriate elevation drawing, the checker indicates this by applying a light-green tone over the E.

Remark: This same technique of using one notation to indicate a multitude of checking operations is also applicable when the checker wishes to indicate that a building elevation drawing needs to be checked against the plans or sections. By placing the notations P and S at one point on an elevation drawing, he avoids having to place those notations on each of the many items which appear on the elevation. As discussed above, some precision might be lost, but considerable time could be saved.

9-8 CHECKING CONSULTING ENGINEER'S DRAWINGS

When the checker is checking drawings made by consulting engineers or other consultants, he analyzes each item for its effect on the architectural drawings or on other engineering drawings, in order to coordinate the physical dovetailing of items in all divisions of the work. The checker is not ordinarily concerned with the correctness of items which do not come within his area of expertise, but merely accepts them with the understanding that they will be checked by the proper engineering personnel. Examples are the sizes of reinforcing bars, model numbers and capacities of HVAC equipment, and sizes of electrical wiring. The sizes of beams, or of HVAC or other equipment, concern the checker only as they relate to other elements of the building; for example, a beam that is too deep to allow a duct to pass under it would obviously be of concern to the checker.

Thus, where the checker encounters the indication of such items as reinforcing sizes, etc. on the check-prints, he applies a light-green tone over the indications to signal that they are of no further concern to him, but are still subject to checking by others.

9-9 REMOVING A SHEET FROM MASTER CHECK-PRINTS

To facilitate the single-sheet check technique, the print which is to be checked should be temporarily removed from the set of master check-prints and placed in a convenient location on a drawing board. The sheet could be folded if desired.

10 TWO-SHEET CROSS-CHECK AND MULTIPLE-SHEET CROSS-CHECK

10-1 TWO-SHEET CROSS-CHECK

Another basic concept of this system of checking is a technique for a complete cross-check of two sheets of drawings against each other, without any interruptions to refer to any other sheets.

A typical sheet that can be efficiently checked by the **two-sheet cross-check** technique is an architectural floor plan, which may be checked against corresponding structural, HVAC, plumbing, and electrical floor plans. This means a cross-check of all items, notes, etc., which does not include the checking of the conformation of spaces and the locations of doors, windows, etc. (which should be checked by using the floor plan overlay technique; see Chapter 17).

The two-sheet cross-check technique is also efficient for checking other portions of the drawings. For instance, finish schedules may be checked against architectural floor plans, elevations may be checked against floor plans, and so on.

10-2 INITIAL PROCEDURE IN TWO-SHEET CROSS-CHECK

The checker should select two check-prints which have both been checked by the single-sheet check technique, with the red notations and the green color code. The two selected sheets are removed from the set of master check-prints, placed side by side on a drawing board, and arranged to facilitate cross-examination of an item on one sheet with an item on the other. Sheets may be folded if necessary, so that items may be compared more easily.

10-3 EXAMPLE OF TWO-SHEET CROSS-CHECK

The checker begins scanning one of the sheets for the particular notation in red which indicates that an item should be checked against the other sheet. For example, if the checker had selected an architectural plan and the corresponding HVAC plan to cross-check, he would scan the architectural plan for the symbol ╲, which means, "Check this item against HVAC drawings." When he finds this symbol marked on (or near) an item on the architectural plan, the checker looks on the HVAC plan to see if the item was properly shown there.

If the item is shown on the HVAC plan, it should have green pencil already applied to it to indicate that it has been checked by the single-sheet check technique. It should also already have one or more notations in red on the print regarding that item; one of those notations being the symbol o, which means, "Check this item on the architectural drawings."

If the checker finds that the item is shown on the HVAC plan, he would apply a green tone over the symbol ╲ on the architectural plan; this indicates that the symbol has been taken into account, and that it requires no further attention. In like manner, the checker applies a green tone over the symbol o on the HVAC plan, which indicates that this symbol, too, has been taken into account and requires no further attention.

If the item is shown on the HVAC plan, but the way it is shown does not agree with the way it appears on the architectural plan, the checker would try to judge which one is correct and then make the necessary notations in red on (or near) the incorrectly shown item to ensure that the item will be corrected.

If the item shown on the HVAC plan does not agree with the way it appears on the architectural plan and the checker cannot decide which one is correct, he would take steps to clear up the conflict by recording a question in the checker's manual for some other person (such as the job captain or the HVAC consultant) (see Chapter 7).

If the item on the architectural plan should have been shown on the HVAC plan but was not, the checker would make notations in red on the HVAC plan to indicate that the item should be added. He also makes notations indicating on what other drawings the just-added item needs to be checked. For example, the symbols o — / might need to be applied, which indicates that the item just added on the HVAC plan needs to be checked on the architectural, structural, and electrical drawings. The checker then immediately applies green to the just-applied symbol o (indicating that the just-added item needs to be checked on the architectural drawing); the checker applies green to the symbol because he has, of course, just checked it on the architectural drawing. (The checker does not apply green to the other red symbols (— and /) until their implied directives, to check the item on the structural and the electrical plans, have been complied with.)

If an item on the HVAC plan has features which should be shown on the architectural plan but are not, the checker would make the necessary notations in red on the architectural check-print to ensure that it will be brought into agreement with the HVAC check-print. For instance, if the HVAC sheet shows ducts which penetrate floors or walls and they are not shown on the architectural check-print, they should be indicated by means of the proper notations in red on the architectural check-print. In addition, other necessary notations should be applied, such as the notation —, to indicate that the structural drawings should be checked regarding provision for the ducts.

10-4 CONTINUING THE TWO-SHEET CROSS-CHECK

To continue the two-sheet cross-check, the checker repeats the same process of scanning one of the check-prints for the symbol indicating that the other sheet should be checked regarding any item marked with that symbol. Following the examples above, this means that the checker continues to scan the architectural plans for the symbol \ on (or near) an item, and when he finds such a symbol he checks the item marked with that symbol against the HVAC check-print, makes any necessary notations, and so forth, as described above.

As the checker proceeds, he applies green to each of the red symbols \ on the architectural check-print to indicate that the symbol has been taken into account, and continues to the point where all indications of that symbol have been taken into account and have green applied to them. The checker then reverses the procedure and scans the HVAC check-print for the symbol o (still in red with no green tone over it) on any item, indicating that the item still needs to be checked against the architectural check-print. When the checker has taken all of the symbols o on the HVAC check-print into account and has applied green to them, the two-sheet cross-check of these architectural and HVAC check-prints is complete.

10-5 TWO-SHEET CROSS-CHECK WITHOUT NOTATIONS

In addition to the technique described above, in which the checker utilizes checker's notations to determine what points on two check-prints need to be cross-checked against each other, the checker may use another two-sheet cross-check technique in certain instances.

For example, the checker cross-checks elevations against floor plans to see that exterior openings and other features agree on the elevations and plans. No checker's notations are needed in this case; the checker simply compares each opening (or other feature) on an elevation with the same item on the floor plan. If they are in agreement, the checker applies light-green pencil around the outline of the item on the elevation, and also on the indication of the item on the floor plan. If any problems are detected, the checker uses checker's notations on the check-prints in conjunction with entries in the checker's manual, to ensure that the problems will be resolved.

10-6 ADVANTAGES OF TWO-SHEET CROSS-CHECK TECHNIQUE

The two-sheet cross-check technique greatly reduces dependence on one of the most inefficient and time-wasting procedures in the usual method of checking drawings, in which the checker selects an item on one sheet and then hunts through the entire check-set for each sheet on which that item should appear in order to check the item on each of the sheets. This hunting through the check-set must be repeated for each item on each sheet of the drawings, and the tremendous number of items involved muliplied by the time required to hunt through the check-set for each of the items can represent a huge amount of time wasted in repetitious activity. The two-sheet cross-check technique of performing an uninterrupted cross-check of one sheet against one other sheet eliminates this repetition; the checker can cross-check dozens of items on the two sheets without having to spend time in repeatedly hunting through the check-set for various sheets on which to check each of the items.

In addition, the two-sheet cross-check technique (in the same manner as the single-sheet check technique) is not dependent on having a complete set of prints. All the checker needs to coordinate sheets are the two sheets he wishes to cross-check against each other. When other sheets become available, he can then cross-check them against each of the first two sheets in the same way.

10-7 MULTIPLE-SHEET CROSS-CHECK

Another concept of this system of the checking is the **multiple-sheet cross-check** technique, which is used for cross-checking items on three or more sheets of drawings at the same time. Elements of the drawings which are most efficiently checked by this technique are door details, cabinetwork details, wall sections, etc. Door details need to be checked against the floor plan, door schedule, and finish schedule; cabinetwork details need to be checked against the floor plan, sections, and finish schedules; and wall sections need to be checked against the floor plan and the building elevations.

For example, if the checker wants to check the door details, he would remove (from the set of master check-prints) the prints which show the door details, the architectural floor plan, the door schedule, and the finish schedule. The checker should arrange these four check-prints on a drawing board in a manner which facilitates examining an item on one of the check-prints, and then on each of the other check-prints. Sheets may be folded if necessary, so that items may be compared more easily. The checker proceeds to check door details in the manner described below (see Step 12 in Chapter 20).

In like manner, any details could be checked most efficiently by this multiple-sheet cross-check technique whenever it is necessary to refer to three or more sheets to check all the items on the details.

The key to efficiency in the multiple-sheet cross-check technique lies in removing the pertinent sheets from the check-set and arranging them so that each item may be checked on each sheet in turn, in a rapid and methodical way, rather than repeatedly hunting through the entire check-set for the proper sheets on which to check each item.

11 COMMUNICATING BY COPY

11-1 COMMUNICATING BY COPY

Another basic concept of this system is that the checker does all the communication of his comments, sketches, lists of details needed, and notations to various persons who will use them, by transmitting copies. This technique allows the checker to communicate his comments, etc., in an efficient way and at the same time to retain his own copy of all the information which has been transmitted.

11-2 GENERAL

The checker communicates his comments, sketches, lists of details needed, and notations to those who will be involved in making the revisions by transmitting copies along with duplicate check-prints on which he has forwarded pertinent notations. The checker retains the originals of his comments, etc., and also retains the master check-prints, which carry all of his notations in red pencil. (The master check-prints and the duplicate check-prints referred to above are described in Chapter 2. It is important that the numbering and arrangement of the check-prints be as described in Chapter 2.)

11-3 FORWARDING PERTINENT NOTATIONS

Forwarding pertinent red pencil notations from the master check-prints to the duplicate check-prints is facilitated by the concise nature of the notation system in general, and also by the fact that only those notations which are directed to the persons responsible for making the revisions have to be forwarded—all those notations which serve as reminders to the checker of actions which he himself must take do not have to be forwarded to the duplicate check-prints.

11-4 NOTATIONS AND OTHER ITEMS WHICH MUST BE FORWARDED

Specifically, the notations which must be forwarded to the duplicate check-prints are the following:

1. Notations which indicate that the checker has recorded (in the checker's manual) a comment, a sketch, or a detail-needed listing. Examples of such notations are AC32, AS16, AD12; SC4, SS10, SD15; HC11, HS9, HD14; PC42, PS17, PD24; EC9, ES19, and ED6.
2. Miscellaneous symbols, such as X˙ and ↰. (The meaning of these notations is described in Chapter 4.)

11-5 DIRECT INDICATION TO BE FORWARDED

In addition to these notations, the checker must also forward to the duplicate check-prints any direct indications of minor changes (see Chapter 4).

11-6 YELLOW PENCIL TONE OVER RED NOTATIONS

A yellow pencil tone is applied over each of the red notations on the master check-print after it has been copied onto the duplicate check-print for transmittal to others (see Chapter 3). This yellow tone serves as an aid in ensuring that all of the red notations on the master check-prints are copied and transmitted, and that none are missed. Also, the yellow tone over the red notations graphically indicates in a most positive way which notations have been copied and transmitted up to that time. Thus, notations which are added subsequently may be readily differentiated from those which have already been copied and transmitted.

This yellow tone should be applied to all direct indications of minor changes which are shown in red pencil on the master check-prints and are forwarded to the duplicate check-prints. In this way, the checker is provided with the same graphic signal as with the forwarded notations, indicating which notations have been copied and transmitted and which ones have not.

11-7 COPIES OF COMMENTS AND SKETCHES

The pages of comments and sketches related to the notations in red forwarded onto the duplicate check-prints may be easily removed from the checker's manual and copied on an office copier for transmittal along with the duplicate check-prints.

The ease with which the checker's comments and sketches may be copied and transmitted is a major advantage of this system of communicating comments by copy. Whereas the notations in red are purposely made concise so that they may be easily copied by hand onto the duplicate check-prints, the checker's comments and sketches may be as extensive as desired because they may be copied on an office copier. The copiers are fast and efficient even when a large number of pages of comments and sketches are involved.

11-8 TRANSMITTING COMMENTS AND OTHER ENTRIES

The comments, sketches, etc., may be transmitted periodically as the checking progresses. The checker should note on each page to be copied the date it is copied, and also the name of the person to whom it is to be transmitted. (On an incomplete page, the checker should draw a cutoff line under the last entry or last sketch and should note under the cutoff line the date and the name of the person to whom the copy has been transmitted, in the same manner as with a complete page.) These indications of which pages (and portions of pages) of comments, sketches, etc., have been copied for transmittal allow the checker to later make copies of additional comments, sketches, etc., without confusing the new material with previously copied and transmitted material.

11-9 COMMUNICATION SUBSEQUENT TO INITIAL COMMUNICATION

After notations and direct indications have been copied onto a duplicate check-print and transmitted to another person, along with copies of comments and sketches, subsequent notations, direct indications, and copies of comments are transmitted in the same manner, with the same procedures to indicate that they have been transmitted. The checker has the option of copying additional notations and direct indications onto a new duplicate check-print, or he could have the first duplicate print returned to him long enough for him to copy additional notations and direct indications onto it.

11-10 JOB CAPTAINS AS INTERMEDIARIES

In general, on a project of any appreciable size, the checker should transmit red-marked duplicate check-prints and copies of related comments, etc., to the job captain (or captains) who is responsible for coordinating the work of the drafter performing the actual revisions. While the checker's comments, etc., are basically directed to the drafter, the job captain needs to be aware of the revisions, at least in a general way. Also, the job captain, rather than the checker, is usually the one to allocate the work to the drafter.

After receiving the duplicate check-prints and copies of comments, etc., the job captain briefly reviews them with the drafter, and the drafter then proceeds with the revisions, referring any questions which arise to the job captain. If some question arose which neither the drafter nor the job captain could answer, they would record the question to present to the checker. The drafter should go as far as possible with the revisions, and then the checker should be called in to discuss the questions at a conference either with the drafter, with the job captain, or preferably with both. (Conferences held to discuss the revisions are described in Article 11-13.)

When the job captain acts as intermediary, to receive and review the checker's directives regarding revisions and to confer with the drafter performing the revisions, the checker has the advantage of a second opinion regarding revisions. Furthermore, this second opinion comes from a person who is close to the job and whose duties presumably include being aware of revisions. If the checker communicates directly with the drafter, the checker would probably have to allocate the work and answer all the drafter's questions, which would have the effect of making the checker assume the responsibilities of the job captain. This might be all right on small projects, but it would be inefficient on large projects.

It should be noted here that the checker, in a way, communicates with the job captain in two ways. One way is through the drafter; many of the comments and questions related to the notations on the duplicate check-prints are directed, in a sense, to both the drafter and the job captain. In making the revisions, the drafter first deals with the checker's comments and questions himself; then if he is unable to make a decision, the drafter would take it up with the job captain. Thus, the job captain receives questions from the checker in an indirect way—through the drafter. In addition, the checker makes entries of questions (in one of the question sections of the checker's manual) which are directed specifically to the job captain. (Entries in the question sections are recorded as described in Chapter 7.) The response to these questions is generally obtained at a checker's conference between the checker and the job captain (see Chapter 15).

11-11 COMMUNICATING COMMENTS BY COPY TO ENGINEERING PERSONNEL

Communicating with engineering personnel is basically the same as communicating with architectural personnel. Here again, the checker usually transmits the duplicate check-prints and copies of his comments, etc., to the engineering job captain (or perhaps to the project engineer for that particular project in the engineering office). It is probable that the checker will not have much verbal communication with engineering drafters, and that the engineering job captain, or perhaps the project engineer, will be the only one with whom the checker will need to confer regarding revisions of the engineering drawings.

After receiving the duplicate check-prints and copies of comments, etc., the engineering job captain or the project engineer and the engineering drafters confer and make revisions in exactly the same manner as the architectural job captains and drafters.

As with architectural job captains, the checker communicates with engineering job captains in the same two ways: through the comments and questions related to the notations on the duplicate check-prints and by entries in the various question sections of the checker's manual (directed specifically to each of the engineering job captains). Similarly, the questions directed to the engineering job captains are answered at checker's conferences.

11-12 INITIAL CONFERENCE REGARDING METHODS

A short initial conference should be held between the checker and each job captain and drafter involved in making the revisions to review the general methods used by the checker, and also to discuss the method of response to be used by the person making the revisions (see Chapter 12). The initial conference should be kept brief and should not involve discussion of details about the comments, etc.; any discussion of details should be deferred until the drafter making the revisions has progressed as far as possible.

11-13 SUBSEQUENT CONFERENCES REGARDING REVISIONS

After receiving copies of the checker's comments and sketches, along with the marked duplicate check-prints, the drafter (aided by the job captain) proceeds as far as possible with the revisions before calling for a conference with the checker to review questions which have arisen regarding the comments.

The drafter should record his own comments in a systematic way as he proceeds; then, a conference concerning the recorded comments could be held. A conference at that time is much more efficient than one held earlier, before the drafter has a chance to become familiar with the problems.

11-14 CONFERENCE NOTES

Adequate notes should be taken by the checker at all conferences, and these notes should be placed in the conference notes file. The taking of conference notes, filing them, etc., will be described in Chapter 15.

11-15 CHECKER RETAINS MASTER CHECK-PRINTS AND MASTER COPY OF COMMENTS AND SKETCHES

A major advantage of communicating comments, etc., by copy is that the checker retains the master check-prints and also the master copy (originals) of his comments, sketches, etc., at all times. This is in contrast to the usual procedure, in which the checker sends his only marked check-prints (often with voluminous comments and sketches marked in red directly on the prints) to the architectural drafting room or to the engineers, thus leaving the checker with a great deal of information unavailable for use in further checking. In addition, the risk of losing a check-print on which the checker has spent a great deal of time and of having no other record of his work is avoided by using this system of communicating comments by copy. If a duplicate check-print is lost in the drafting room, the checker can easily prepare another one from the master check-print.

11-16 FLEXIBILITY

Another advantage of this system of communication is that the checker may partially check a print, and at any time he may stop and transmit his notations, comments, sketches, etc., to whomever he wishes. Then he can resume his checking until he again desires to transmit further notations, comments, etc.

There are no problems with this transmittal of notations, comments, etc., in increments, because the checker knows at all times what information has been transmitted and what information has not yet been transmitted. This is readily determined by observing which of the red notations on the master check-prints have yellow applied over them, indicating that they have been transmitted, and by observing the cutoff lines and dates on pages in the checker's manual, which indicate which comments, etc., have been transmitted.

This flexibility aids in the efficient utilization of the personnel making the revisions. The checker can start someone on making revisions on a sheet without having to check the sheet completely. After starting someone on revisions, the checker could then return to his checking, since he retains at all times a complete record of his notations, comments, sketches, etc.

12 REVISING THE DRAWINGS

12-1 REVISING THE DRAWINGS

Another concept of this system of checking is that the architectural and engineering drafters and job captains involved in making the revisions use certain techniques similar to those employed by the checker. These techniques are graphic in nature and greatly facilitate the efficiency of making the revisions.

12-2 JOB CAPTAIN AS INTERMEDIARY

The checker transmits the red-marked duplicate check-prints and copies of his comments and sketches to the job captain who will be responsible for the revisions of those particular drawings. The job captain acts as an intermediary for the checker by allocating the work to the drafter and also by helping the drafter interpret the intent of the checker's comments and sketches (see Chapter 11).

The job captain should give one sheet of the duplicate check-prints along with copy of checker's comments and sketches to each drafter who will be making revisions; the job captain then reviews with each drafter the specific duplicate check-print and the copies of related comments and sketches that he has been given. The drafter then proceeds with the revisions, using the techniques described below.

12-3 DRAFTER'S COLOR CODE

A drafter making revisions uses a **drafter's color code**, similar to the checker's color code, for marking the duplicate check-print which he has been given. The drafter's color code is detailed below.

12-4 LIGHT GREEN

As the drafter makes each revision in response to the checker's comments and sketches, he should apply light-green pencil over the red notation related to that particular revision. The light-green tone over the red notation signifies that the notation has been taken into account and requires no further attention. In this way, the drafter is assured that he has not missed any of the notations; if a red notation does not have light green applied over it, it still needs attention. (Note that this use of light green is similar in significance to the checker's color code described in Chapter 3.)

12-5 ORANGE

The drafter uses orange pencil on the duplicate check-prints to indicate several things. Orange pencil is used in order to differentiate the drafter's notations from the checker's notations, which are in red pencil. Eberhard-Faber Colorbrite Orange #2122 is recommended for the drafter's marking of the duplicate check-prints. If the drafter does not understand the intent of one of the checker's comments or sketches, he would use orange pencil to circle the related checker's notation on the duplicate check-print. The orange circle serves as a flag to indicate that the drafter needs to have the related comment or sketch clarified by the checker. Also, if the drafter, for some reason, makes a revision which he knows is not in accord with a comment or a sketch made by the checker, he would again indicate this by applying an orange circle around the related checker's notation on the duplicate check-print. The checker should be advised of the drafter's reasoning at a subsequent conference. If the drafter initiates a revision himself, or makes a revision initiated by a job captain or other person, he would apply orange pencil on the duplicate check-print to indicate this to the checker.

12-6 DRAFTER'S COMMUNICATIONS

If the drafter needs to communicate other information to the checker, he would use orange pencil on the duplicate check-print to indicate brief comments or questions. If the comments or questions are lengthy, the drafter should record them in a loose-leaf binder similar to the checker's manual. The drafter could employ a simple numerical notation system to tie such comments and questions to points on the drawings. To do this, the drafter places a number in orange on a check-print, that number being the same as the number of the written entry in his notes. Sketches by the drafter could be handled in the same way.

The system of communication used by the drafter is similar to that used by the checker, but does not need to be at all elaborate, as the drafter needs to communicate with only one person—the checker.

12-7 DETAILS NEEDED

In addition to the copy of the checker's comments and sketches which the drafter uses, the job captain makes use of a copy of the list of details needed that the checker has recorded in the checker's manual (see Chapter 6). The job captain uses this list in assigning work to various drafters. If the job captain or any of the various drafters have unresolved questions about this list of details needed, they should discuss these problems with the checker.

12-8 AFTER THE REVISIONS HAVE BEEN COMPLETED

After the revisions of a drawing have been completed, the drafter has three new check-prints made, and he or the job captain transmits two of these to the checker, retaining one for drafting-room reference. At the same time, the drafter or the job captain returns the duplicate check-print of that drawing, including all the red checker's notations, now covered with light-green pencil by the drafter (perhaps with some notations or revisions circled in orange by the drafter as well). The revisions are then verified by the checker (see Chapter 13).

12-9 STRUCTURAL, HVAC, PLUMBING, AND ELECTRICAL REVISIONS

The procedures described above apply equally to revising architectural, structural, HVAC, plumbing, and electrical drawings.

13 CHECKER'S VERIFICATION OF REVISIONS

13-1 CHECKER'S VERIFICATION OF REVISIONS

Another concept of this system of checking is the technique whereby the checker verifies that the drawings have been properly revised in response to his directives. This technique will be described in the following articles.

13-2 NEW CHECK-PRINTS OF REVISED DRAWINGS

After the revisions of a drawing have been completed, the drafter making the revisions has three new check-prints made, and he or the job captain transmits two of these to the checker, retaining one for drafting-room reference (see Chapter 12). At the same time, the drafter or the job captain returns the duplicate check-print of that drawing, including all the red checker's notations, now covered with light-green pencil by the drafter (perhaps with some notations or revisions circled in orange by the drafter, as described in Chapter 12).

13-3 DATING, DESIGNATING AND ISSUE-NUMBERING OF NEW CHECK-PRINTS

The checker uses the procedures described in Chapter 2 to apply a date, a designation, and an issue-number to each of the two new check-prints. In this manner, one of the new check-prints receives the designation MCP # _____ (master check-print # _____), and the other new check-print receives the designation DCP # _____ (duplicate check-print # ____). The number in each case is the issue-number to indicate the numerical sequence of the new check-prints. If this is the third set of check-prints to be made from that drawing, one of the check-prints would be indicated as MCP #3 and the other would be indicated as DCP #3. The date indicated on each check-print is the date they were received by the checker.

When the checker establishes an issue-number for the latest check-prints, he should advise the person who retained the third check-print for drafting-room reference. That person should designate the drafting-room check-print with the initials DRCP (drafting-room check-print), plus the same date and issue-number as the other two check-prints.

The dates, designations, and issue-numbers on all check-prints should be indicated in Colorbrite Green #2128 (see Chapter 2).

13-4 CHECKER'S PROCEDURE IN VERIFYING REVISIONS

The checker first lays aside the new check-print designated DCP # ____ for use later in communicating any further notations, along with comments and sketches. Next, the checker takes the other new check-print, designated MCP # ____ and places it on a drawing board. Then, beside this new master check-print, the checker places the previous master check-print of that same drawing. The previous master check-print has the checker's various notations on it, applied in red pencil. Those notations which have been communicated to the drafters making the revisions have a yellow tone applied over the red notation, resulting in a yellow/red tone (see Chapter 3). The checker then selects one of these yellow/red notations and looks in the checker's manual to ascertain which comment, sketch, or listing of a detail needed it refers to.

The checker then examines the new master check-print to determine if it has been revised in accord with the intent of the comment or sketch he recorded; or if an additional detail was called for, whether the detail now appears on the new master check-print.

If the checker finds that the new master check-print has been properly revised to reflect the intent of the yellow/red notation on the previous master check-print and its related entry in the checker's manual, he would then apply a light-green tone over the yellow/red indication on the previous master check-print, which indicates that it has been taken into account and requires no further attention.

As the checker reviews his entries in the checker's manual to verify that the revisions have been properly made, he applies a light-green pencil tone over all elements of his comments, sketches, etc., with which the revisions agree. He also applies a green vertical line through each entry when he has verified that the related revision has been properly made. (This technique of signaling that comments, sketches, etc. have been taken into account by means of the light-green tone and the vertical green line is described in Article 8-2.)

13-5 COMPLETING THE VERIFICATION OF REVISIONS

The checker should continue in this manner, taking each yellow/red notation on the earlier master check-print and verifying that the indicated revisions have been properly made; and then applying light-green pencil over each yellow/red notation to indicate that it has been taken into account and requires no further attention. When all the yellow/red notations have had light green applied over them, the checker is through with verifying the revisions made on that sheet.

13-6 REVISION IMPROPERLY MADE

If the checker finds that a revision has not been made in accord with his comments or sketches, or that a needed revision has not been made at all, he would not apply green over the related yellow/red notation on the previous master check-print but would leave it as it is. This acts as a flag to call his attention to the fact that the notation has not been properly taken into account. The checker then goes to the new duplicate check-print and, again, forwards the notation onto that new print in red pencil. This new duplicate check-print is later sent to the drafter, who is thus reminded of the desired revision as indicated by the "reappeared" notation in red. Only when the checker finally verifies that the revision has been properly made does he apply green on the previous master check-print over the yellow/red notation related to that revision.

13-7 DRAFTER'S INDICATIONS IN ORANGE

As he verifies each revision, the checker also checks on the returned duplicate check-print to see whether the drafter has made any indications (in orange pencil) related to that revision. Also, the checker observes any markings in orange pencil which indicate that the drafter has made a revision on his own initiative. The checker takes these drafter's indications into account as he proceeds with his verification of the revisions (applying light green over each orange indication).

13-8 FURTHER COMMUNICATIONS WITH PERSONS MAKING THE REVISIONS

As the checker proceeds with his verification of the revisions, if any problems arise which he wishes to communicate to the persons making the revisions, he would proceed according to the procedures described in Chapter 11.

Further revisions are made by the drafter in accord with the latest communications from the checker, and then the checker verifies those new revisions. This process is repeated as required, until all revisions have been made and verified.

14 ASSISTANT CHECKERS AND SELECTIVE CHECKING

14-1 ASSISTANT CHECKERS

Another concept of this system of checking is a technique whereby the checker may utilize assistant checkers and yet retain complete control of the overall checking process, ensuring that the checking will be comprehensive and completely coordinated.

First, the checker assigns portions of the checking to one or more assistants. This is done by printing extra check-prints of the drawings as required for the assistants to do their checking. These extra check-prints are each designated by the initials ACP (assistant's check-print), along with the same date and issue-number as the checker's master check-prints.

For example, assume that a check-print of sheet A-9 was printed for an assistant checker's use. At the same time, the checker makes a check-print to keep his master check-prints up to date. If we assume that the issue-number of the checker's print was MCP #3, then the assistant's check-print would carry the designation ACP #3, indicating that it was an assistant's check-print and that it was a duplicate of master check-print #3. In addition, the date on which the check-prints are made should be indicated. Both the designation ACP #3 and the date are applied in Colorbrite Green #2128.

After they have received the necessary assistant's check-prints, assistant checkers proceed with their checking in the same manner as the checker, except that the assistants limit themselves to the particular area of the work assigned to them by the chief checker. Assistant checkers employ all of the techniques used by the chief checker, including the checker's color code and the checker's notations. In addition, each assistant checker provides himself with an assistant checker's manual, to record his comments, sketches, etc., which is used in exactly the same way as the checker's manual.

There are basically two types of assignments which the chief checker might give to an assistant. One type is an assignment of a large block of work, such as checking the door details shown on a certain drawing; the other type is the checking of items which are scattered throughout the drawings, such as all the dimensions on a project.

Assume, for example, that the checker assigns the checking of door details on a certain drawing to an assistant. To do this, the checker takes the master check-print of that particular drawing and circles the entire area of the door details in Colorbrite Green #2128, to indicate that they have been assigned to an assistant. In addition, the checker adds a note in Colorbrite Medium Red #2126 on the master check-print stating, "Door details being checked by ____________" (naming the assistant checker).

Again, let us assume, for example, that the checker assigns the checking of all dimensions on a project to an assistant. To do this, the checker adds a note in red pencil, on all sheets of the master check-prints which are involved, stating, "Dimensions being checked by ____________" (naming the assistant checker). Then, as he continues with his own checking, the chief checker applies light green to each indication of a dimension.

The circling in green of the block of door details by the checker does not mean that the door details are necessarily correct, but it does indicate that the chief checker is accepting them provisionally, subject to any corrections which the assistant checker might make. The same is true when the chief checker applies light green to the scattered items, such as dimensions, which were assigned to an assistant checker to check.

14-2 COMMUNICATING ASSISTANT CHECKER'S COMMENTS, SKETCHES, AND NOTATIONS

Assistant checkers communicate with persons making the revisions in the same manner as the checker (see Chapter 11): by providing the person making revisions with a copy of notations in red on a duplicate assistant's check-print, along with copies of the assistant checker's comments and sketches recorded in the assistant checker's manual. Assistant checkers also verify the correctness of revisions in the same manner as the checker (see Chapter 13).

14-3 ALTERNATE METHOD OF COMMUNICATING ASSISTANT CHECKER'S COMMENTS, SKETCHES, AND NOTATIONS

If the chief checker chose to, he could have the assistant checkers turn their marked assistant's check-prints and their comments and sketches over to him, and he (the chief checker) could proceed to communicate with the persons doing the revisions and to verify the correctness of the revisions.

This method has the advantage of familiarizing the chief checker with the revisions of areas of the work assigned to assistants, and could perhaps prevent the confusion which might be caused by having several different checkers communicating with the persons doing the revisions. A disadvantage is that the chief checker might be fully occupied with other areas of the checking, and therefore it would not be practicable for him to assume the task of communicating assistant checkers' comments, etc. Another disadvantage is that the assistant checkers are more conversant with exactly what revisions are required. Thus, if they communicate directly with the persons doing the revisions, it might be more accurate than having the chief checker communicate directives which are necessarily somewhat secondhand to him.

The decision about who should handle the assistant checker's communications has to be made by the chief checker according to each particular project.

14-4 SELECTIVE CHECKING

By using this graphic system of checking, the checker has complete flexibility as to what portions of the drawings he chooses to check. As discussed previously, the checker may choose to check a certain portion of the drawings in order to most efficiently utilize personnel. If the architectural personnel are getting caught up, he might wish to concentrate on checking architectural drawings. In like manner, he might elect to check structural, HVAC, plumbing, or electrical drawings, according to his judgment as to what would most facilitate progress.

In addition to aiding in the efficient utilization of personnel, **selective checking** enables the checker, in an emergency situation, to partially check a set of drawings prior to issuing the drawings, and to then go back, finish the checking, and issue any necessary addenda and revised drawings during the bidding period.

By using selective checking, the checker can concentrate on the items which he considers to be the most important prior to releasing the drawings for bids. Items of most importance to bidders are such things as the major components of the structural, mechanical, and electrical systems; the major exterior materials; interior finishes; door and frame materials; types of windows; and major dimensions, such as overall dimensions and floor-to-floor dimensions. The checker could go through the drawings and try to check as many of these major items as possible, and then, after the drawings were issued, he could go back and check the other items. The graphic nature of this system of checking allows him to go back and later check the less important items, without having to recheck all of the previously checked items.

While such partial checking is not by any means the best way to proceed, it is obviously much better than issuing drawings for bids without any checking at all; and the technique of selectively checking the most important items is also better than a hit-or-miss partial check.

15 CHECKER'S CONFERENCES

15-1 CHECKER'S CONFERENCES

In the course of the checking, conferences between the checker and various other persons are necessary. Such conferences are called **checker's conferences**.

In this chapter, a number of concepts will be presented which will aid the checker in making checker's conferences as productive as possible, ensuring that the checker will be able to get all his questions answered expeditiously and keep a complete and orderly record of each conference.

15-2 GENERAL

In addition to the conferences described previously between the checker and the various job captains and drafters involved in revising the drawings, the checker needs to have conferences with the various persons to be asked questions (questions recorded in the question sections of the checker's manual, as described in Chapter 7).

A checker's conference might be very brief and informal, with only a few questions involved; or it might be longer, more formal, and have a large number of people in attendance. Brief, less formal conferences are referred to as **minor conferences**, and longer, more formal conferences are referred to as **major conferences**. Both types of conferences are handled in the same basic way, but there are certain differences.

The frequency of checker's conferences depends on how fast the checker accumulates questions he needs to get answered. It is obviously more efficient to get answers to a fairly large number of questions than it is to arrange a conference to answer just one or two questions. However, the checker should not delay too long in getting answers, because some of the answers might affect ongoing work, and having the answers on a reasonably current basis might save time in the long run.

15-3 CONFERENCE PRINTS, GENERAL

At all checker's conferences, there is almost always a need for prints of the drawings, both for reference and also for applying marks and notations during the conference. *The checker should avoid using his master check-prints at a conference.* The checker's notations on the master check-prints should not be confused by the addition of marks and notations made at a conference.

Conference prints are divided into two types: those used for minor conferences and those used for major conferences. These two types of conference prints will be described in the following articles.

15-4 MINOR CONFERENCE PRINTS

The checker should maintain one set of reasonably up-to-date prints to be used on a continuing basis for the short, intermittent minor conferences which occur on a day-to-day basis. This set of prints is called the **minor conference prints**.

As work on various sheets of the drawings progresses, the checker should make progress prints and insert these in the set of minor conference prints. Overall, this set should be kept in normal numerical order; all successive progress prints of the same sheet should be included and be arranged in reverse chronological order, with the latest print on top. The set of minor conference prints should be held together with Plan-Hold friction binders, or with a number of spring-type clamps (or another, similar product).

15-5 INACTIVE MINOR CONFERENCE PRINTS

As soon as possible after each minor conference, the checker reviews all the marks and notations made on the minor conference prints and takes actions as required in response to them. The checker applies light-green pencil on the prints over the marks and notations, as a signal that they have been taken into account.

In the meantime, the checker might have another minor conference and apply more marks and notations, as a more or less ongoing process; also, more work might be proceeding on the tracings from which the prints were made. After a particular minor conference print has accumulated a considerable number of marks and notations on it, or a considerable amount of additional work has been done on the tracing for that print, the checker makes a new progress print of that sheet and places this new print in the set of minor conference prints. Then, when all of the marks and notations on the earlier print have been taken into account and have green applied to them, the checker removes that earlier print from the set of minor conference prints and places it in a separate inactive set, called the **inactive minor conference prints**.

The inactive minor conference prints are kept in the same order as the active minor conference prints, as described above. Also, this inactive set is held together in the same way as the active set. The inactive prints should be retained by the checker until such time as he is reasonably sure they are no longer needed.

15-6 MAJOR CONFERENCE PRINTS

In addition to maintaining the set of minor conference prints, the checker should provide an additional fresh set of prints for each major conference.

It is obvious that if a conference is one of major importance, it would be advisable to provide a set of prints made expressly for that conference. If this is done, it would allow the marks and notations made on the prints at a major conference to be entirely separate from the marks and notations made at any other conferences. This can be important when decisions made at one conference supersede decisions made at an earlier conference. In addition, the visual confusion of adding marks and notations made at a major conference on top of marks and notations made at previous conferences should be avoided, if possible.

Major conference prints should be stapled in a set, and the set should not be taken apart, either at the conference or after the conference. The stapled sets of major conference prints should be kept in reverse chronological order according to the conference date on each set, with the latest set on top. Plan-Hold friction binders should be used to hold all the various sets together.

15-7 INACTIVE MAJOR CONFERENCE PRINTS

The checker should maintain sets of **inactive major conference prints** in a manner similar to that described above for the inactive minor conference prints. The only difference is that, unlike the inactive minor conference prints, which are retired to inactive status on a sheet-by-sheet basis, the inactive major conference prints are retired to inactive status on a set-by-set basis. Therefore, the sets of major conference prints should not be taken apart, and a set is not retired to inactive status until all the marks and notations on every sheet in that set have been taken into account.

Like the active sets of major conference prints, the sets of inactive major conference prints are kept in reverse chronological order, with the latest set on top. They are placed in Plan-Hold friction binders, and should be retained by the checker until he is reasonably sure they are no longer needed.

15-8 AVOIDING EXTRA SETS OF MAJOR CONFERENCE PRINTS

In general, it is recommended that there be only one set of conference prints at a checker's conference. If extra sets of prints are furnished to other persons at the conference, there is a tendency for the conference to disintegrate into "splinter groups," with each group discussing different things and marking the various sets of prints in different ways. Having only one set of prints aids the checker in focusing the conference on answering his recorded questions and in keeping the conference organized so that it proceeds efficiently. If this is not done, serious problems can ensue when the answers to questions, or other comments from the conference, are recorded.

Having only one set of prints at a conference means that the checker is responsible for reviewing the conference decisions, as indicated by notations on the prints and by his conference notes, and relaying the decisions to various persons, including those persons who will be revising the drawings. (The checker does this in the manner described in Article 15-22.)

15-9 DATES ON CONFERENCE PRINTS

The checker should place a date on all conference prints, whether they are minor or major conference prints. This date should be preferably in dark-green colored pencil, such as Eberhard-Faber Colorbrite #2128.

On minor conference prints, the date is the date the print was received by the checker; this determines the chronological order of succeeding prints of the same drawing. On major conference prints, the date (preferably on each sheet of the set of prints) is the date of the conference for which the prints were made; this determines the chronological order of the sets of major conference prints, which would then determine the order in which the sets were filed. In addition, major conference prints should carry the date on which they were printed.

15-10 ITEMS NEEDED BY THE CHECKER AT CONFERENCES

The checker should have certain items at conferences to ensure maximum efficiency. If the checker is conferring with a person or persons to whom he has addressed questions, the checker obviously would have the question sections of the checker's manual on hand. If the checker has elected to maintain a separate binder for the question sections, he would bring it to the conference. If the checker kept the question sections in the same binder with the rest of the checker's manual, he would remove those question sections (including all the indexing dividers) and place them in a separate three-ring binder prior to each conference. For efficiency, the checker should *not* take the entire checker's manual to conferences.

In addition to a binder containing the question sections, the checker should have a vinyl folder which has an 8½ × 11 in. wide-ruled note pad (punched for use in a three-ring binder) on one side, and a pocket on the other side to hold completed pages of notes, which are removed from the pad as the conference progresses.

The checker should have a red ballpoint pen to make conference notes on the note pad, to record answers in the question sections, and to make notations on conference prints. The red color is needed for notations on the prints so that they would not be missed in reviewing the prints later. The same pen should be used for the conference notes made on the note pad and for answers in the question sections, because it is imperative that the checker be able to move very rapidly—making notes on the pad, in the question sections, and on the prints—without having to stop and change pens.

The checker should have a mechanical pencil with a fairly soft lead to make sketches at the conference. Pencil is preferred for sketches because it is erasable. The pencil should be a mechanical type, which does not require sharpening, for purposes of speed. If the pencil does not have an adequate eraser, the checker should bring one or more supplementary erasers of a suitable type, such as the Faber-Castell #1962 Auto Magic-Rub vinyl eraser in a chuck-type holder.

The checker should have a briefcase at any conference away from his office. This briefcase should be adequate to contain the three-ring binder of question sections, the vinyl folder, tracing paper, extra note pads, pencils, lead, erasers, and red ballpoint pens.

The checker generally needs some type of conference prints at each conference. If it is a minor conference, the checker would have a set of minor conference prints; if it is a major conference, the checker would have a set of major conference prints. If other prints are marked with notations at a conference, they should be available for the checker to take with him and include in his conference print files. If this is not possible, the checker should copy any notations onto duplicate prints for his files.

15-11 CONFERENCE NOTES

The checker keeps a complete graphic record of the course of each conference. He records all answers to questions directed to persons at the conference, all further questions and comments which arise during the conference, and all decisions arrived at during the conference. In addition, the checker retains all sketches made by himself or by others during the course of the conference. For convenience, the term **conference notes** will be taken to include sketches, as well as written notes.

As speed is essential in recording conference notes, it is recommended that the checker use longhand for the written notes, with the proviso that legibility be strictly maintained. As described in Article 15-10, the written conference notes should be recorded in red ballpoint ink and sketches should be made in pencil.

Figure 15-1 illustrates how a typical page of conference notes appears and shows some of the various forms which conference notes might take. These forms and other features of Figure 15-1 will be described in the following paragraphs.

At the top of the page, the date is the date of the conference, and the project number is the number of that project. The number 1, in the upper-right corner of the page, is the chronological number of the page, indicating that it is page 1 of the conference notes for this conference. Subsequent pages 2, 3, 4, etc., would be placed on top of page 1, in reverse chronological order, with the latest page on top. (Note that the pages of conference notes should be grouped conference by conference, and that the numbering of the pages starts over for each separate conference.) The CN in the upper-right corner of the page indicates that this is a page of conference notes.

The checker should record the names of all persons attending the conference. As an alternative, the checker could pass a note pad around and have each person sign his name and indicate the organization to which he belongs. This page of names then becomes the first page of the conference notes.

The indications (1), (2), (3), etc., are the chronological numbers of the conference notes entries (and sketches) which record the course of the conference. These entry numbers start with (1) on the first page of each conference and run consecutively through all of the succeeding pages of conference notes from that particular conference. As with the page numbers, these conference note numbers start over with each conference.

Entry (1) indicates that James Stanley (shown by the initials "JS") says that beam 213 is 12 × 24 in. Entry (2) indicates that a door (or doors) on the conference prints has been marked by the checker (in

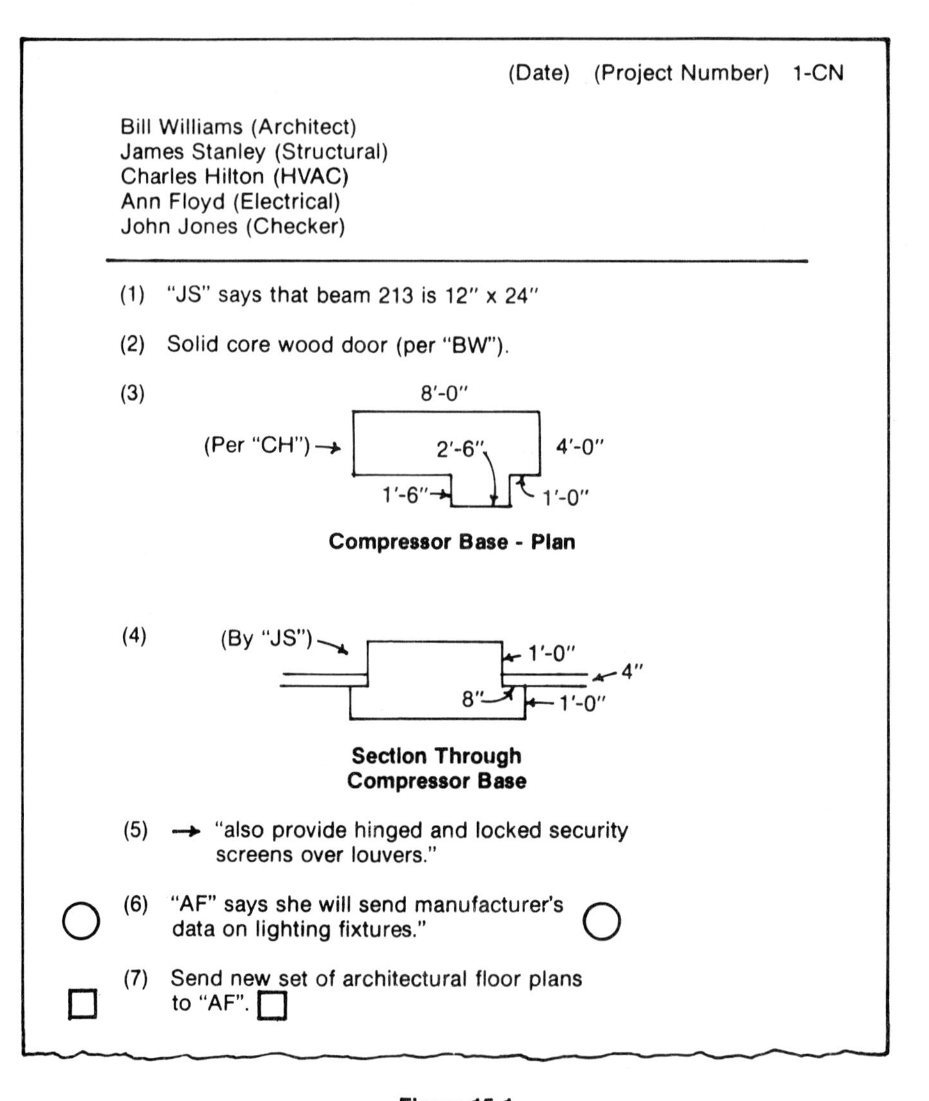

Figure 15-1

red ballpoint ink) with the notation 2, which is called a **conference print notation** and will be described further below. The indication "per BW" means that Bill Williams advised the checker that doors so marked should be solid-core wood doors.

Entry number (3) is a sketch made by the checker. The "per CH" indicates that this sketch is in accord with information supplied by Charles Hilton.

Entry (4) is a sketch made by James Stanley, as indicated by the note "by JS" (the checker, rather than James Stanley, indicates the entry number (4) and indicates that the sketch was made by JS; the checker also supplies the title of the sketch if necessary).

Entry (5) represents a partial answer which is a continuation of an answer started in one of the question sections of the checker's manual. The small arrow preceding this entry indicates that this entry is a continuation of the partial answer recorded in the question section. Recording of answers in the question sections, and the continuation of those answers in the conference notes if the answer is very long, will be described in Article 15-13.

Entry (6) illustrates a technique for flagging an action which has been promised by a person other than the checker. In this example, Ann Floyd ("AF") has promised to send some data to the checker. The checker draws a circle immediately after the indication of what has been promised, and he draws another circle in the border to the left of the entry. Both of these circles are drawn with the red ballpoint pen which is being used to make the conference notes, and they serve as an easily recognizable signal to alert the checker that some action has been promised by another person. Later, when the promised action has been taken, the checker applies a checkmark in green in both of the circles, to indicate that they have been taken into account and require no further attention.

Entry (7) illustrates a technique for flagging an action which the checker himself needs to take. Here, the checker is reminding himself to send a new set of architectural floor plan prints to Ann Floyd ("AF"). This time, the checker draws a square immediately after the indication of the action he needs to take, and he draws another square in the border to the left of the answer. Both of the squares are drawn with the red ballpoint pen, and both are instantly recognizable as signaling an action which the checker must take. The checker applies a checkmark in green in both of the squares when he has completed the action called for in this entry.

15-12 ANSWERS TO CHECKER'S QUESTIONS

As described in Chapter 7, the checker records all questions directed to other persons in the question sections of the checker's manual. The checker brings all of the question sections, including all of the indexing dividers between the sections and subsections, to each conference in a three-ring binder.

At the conference, the checker selects any desired question subsection and then proceeds to read each of the questions in that subsection to the person to whom they are directed. Where a question refers to an item on a drawing, the checker uses the drawing location key included with the question to find the related point on the conference prints. After reading a question to the person, the checker records the answer as follows.

The checker records the answer in the one or two lines of space that were left open following the question when it was recorded (see Chapter 7). The checker uses the red ballpoint pen to record the answer. As soon as he has recorded the answer, the checker draws a vertical red line through the question and also through the answer, to indicate that the question has been answered. This vertical red line is drawn at approximately the center of the page, to appear as shown in Plate III. As other questions are answered, the checker draws a similar vertical red line through each question and its answer. All these lines are drawn at approximately the center of the page, so that the segment of red line drawn through one question and answer would connect with the segment drawn through the question and answer above and below. In this way, when all questions have been answered, the connected segments of red line run unbroken from top to bottom of the page. This unbroken red line provides an instantly recognizable signal that all questions on the entire page have been answered.

The vertical red line does not signal that the checker is through with a question and its answer; he still must review the question and answer and take all necessary actions in accord with them. A *green* line (parallel to the red line) will be drawn vertically through each of the questions and answers when the checker has taken the actions required by each question and its answer; this is shown in Plate III and is further described in Article 15-22.

15-13 ANSWERS TOO LONG FOR OPEN SPACE PROVIDED

If an answer is very long, the checker would record as much of it as possible in the one or two lines of open space which he provided after each question and then record the remainder of the answer on the conference notes pad. Then, in order to tie the two separated portions of the answer together, the checker would utilize the conference note number of the second portion of the answer which is recorded in the conference notes.

For example, the checker observes that the last part of an answer will be recorded as conference note number (5), as shown in Figure 15-1. Either before or after recording this last part of the answer, the checker returns to the first part of the answer recorded in the question section, and there, at the very end of the partial answer, he writes the number (5), thus indicating that the continuation of the partial answer can be found at conference note number (5). The checker places a small arrow after the number (5) to further indicate that the entry is continued. The number (5), the partial answer in the space after the question, the small arrow, and the continuation of the answer recorded in the conference notes are all recorded with the same red ballpoint pen described in Article 15-10.

15-14 QUESTIONS OTHER THAN THE CHECKER'S QUESTIONS

If persons other than the checker raise questions during a conference, the checker would record such questions in the conference notes, indicating who asked the question in each case. The checker also records answers to such questions, and indicates who supplied the answer in each case, in the conference notes.

15-15 CONFERENCE PRINT NOTATIONS

As the conference proceeds, if the checker or another person wishes to make a comment, ask a question, or make a sketch which is related to an item on the conference prints, this would be accomplished by use of a technique similar to the checker's notations described in Chapter 4.

Assume that a certain conference print is under discussion at a conference, and that architectural job captain Bill Williams advises the checker that door number 322 on that drawing must be a solid-core door. The checker examines the conference notes to ascertain the number of the last conference note. Let us assume that the previous conference note was number (1). The checker then adds a new conference note (with red ballpoint pen) below the previous conference note as follows:

(2) Solid-core wood door (per "BW")

After making this entry number (2) in the conference notes, the checker then returns to the conference print, and on (or near) door number 322 adds the notation 2 with a red ballpoint pen. (Checker does not use parentheses around such check-print notations; omitting parentheses speeds up the process, and notations occupy less space on the drawing.) This notation 2 on the conference print is called a **conference print notation**, and it indicates that conference note number (2), relating to door number 322, has been recorded by the checker.

It should be pointed out that a conference note concerning an item on a conference print does not necessarily have to be in the form of a comment, as in the example above, but could be in the form of a question or a sketch. Also, in addition to the use of the conference print notation system, very simple directives can be noted directly on the conference prints, but such direct indications should obviously be extremely short and concise.

Remark: It is obvious that the conference print notation system helps the checker avoid a confused clutter of notes and sketches on the conference prints. This is particularly true in the case of repetitious notes or sketches; if a note or a sketch were related to a number of different points on a drawing, the checker would be able to repeat the same conference print notation at the appropriate points, without having to repeat the note or sketch. In the example above, the notation 2 placed on any door on the conference prints indicates that the door should be a solid-core door.

15-16 OTHER PERSONS' NOTES, SKETCHES, AND MARKS ON PRINTS

If other persons at a conference wish to keep notes, and if they make sketches in their own notebooks, the checker should still make his own notes and not depend on their notes or sketches (which may be unavailable to the checker).

If the checker can get other persons to make their notes and sketches with the checker's notes and sketches (on the checker's conference notes pad), it would help in coordinating their notes and sketches with the checker's questions. If this cannot be done, and if the checker needs the other persons' notes and sketches later for his review of the conference, he should (if possible) get copies of their notes and sketches for his files. These copies should be filed along with the checker's conference notes for that conference.

If other persons mark up the conference prints, the checker should add any conference print notations and related conference notes necessary to supplement or clarify the other person's marks. By doing this, the checker can use the other person's marks to the best advantage when he reviews them later.

15-17 SKETCHES ON THE CONFERENCE PRINTS

If possible, sketches should be made only on the conference notes pad and not on the conference prints. The conference prints may be marked with short, simple indications when necessary, but the checker should attempt to keep the conference prints as free from clutter as he can, and keeping the sketches on the conference notes pad will aid in this.

15-18 SKETCHES ON TRACING PAPER

Occasionally, the checker or some other person at a conference needs to make a sketch which necessitates tracing, such as a sketch of an item on the conference prints. In this event, the checker should have some tracing paper on hand (in his briefcase) in perhaps two sizes. A recommended smaller size is 8½ × 11 in., punched for insertion in a three-ring binder.

If any sketches (whether traced or not) larger than 8½ × 11 in. need to be made, it would be advisable to make them on tracing paper so that they can be reproduced (as a diazo blueline print) for transmittal to various persons who might need them later. To make such oversized sketches, the checker should have several pieces of tracing paper of reasonably large size on hand. Sheets which are 36 × 24 in. will fold down to a 9 × 12 in. size, and several of these could be stored in the checker's briefcase.

15-19 CONTINUING THE CONFERENCE

After the checker has finished with all the questions in the first question subsection, he selects another question subsection and proceeds in the same manner. In this way, the checker goes through all of the question subsections that contain questions directed to people at the conference.

As each page of conference notes is filled up, the checker removes that page from the note pad, and places it in the pocket in the vinyl cover (latest page goes on top of the previous pages, so that they are arranged in reverse chronological order).

At the end of the conference, the checker makes the note, "End of Conference (date)," and draws a line horizontally across the page under that note to indicate the termination of the conference.

15-20 TELEPHONE CONFERENCES

Telephone conferences are handled in basically the same way as any other checker's conference. The checker makes conference notes and sketches and cross-references them to the conference prints with conference print notations.

15-21 MISCELLANEOUS CONFERENCES

When the checker has a situation in which someone wants to confer with him and he does not have his conference materials at hand (binder of question sections, pad for conference notes, and conference prints), a somewhat different procedure would be used.

The checker obtains an 8½ × 11 in. note pad on which to make conference notes. He indicates the date of the conference, the project number, and the name(s) of the other person(s) at the conference, on the first page in the usual manner. After that, the conference proceeds in the same manner as any other checker's conference, except that there are not any questions from a question subsection to guide the conference.

As the checker does not have his conference prints, the prints used at the conference, for reference or marking, are someone else's. The checker uses these prints in the same way as he would his own conference prints, marking them and applying conference print notations to tie various points on the prints to related conference notes (and sketches). After the conference, if the checker is not able to take these prints for his files, he would borrow them and forward the conference print notations to duplicate prints, which he should retain.

After a miscellaneous conference, the checker should put the prints which were used as conference prints (or the duplicates as described above) in his set of minor conference prints.

15-22 CHECKER'S PROCEDURE AFTER EACH CONFERENCE

The first thing the checker should do after each checker's conference is to remove all the conference notes pages from the vinyl folder he used at the conference, and place the pages in a loose-leaf binder called the **conference notes file**. The pages are arranged in reverse chronological order, and the group of pages from each conference should be stapled together with the staple in the upper-left corner to provide access to the notes without removing them from the binder.

Next, as soon as possible after each conference, the checker takes the steps necessary to see that all revisions of the drawings and any other actions required as a result of the conference are accomplished. Basically, this means that the checker reviews the conference notes, the question subsections used at the conference, and the marked conference prints; makes the necessary checker's notations on the master check-prints; and enters the necessary related comments or sketches in the appropriate sections of the checker's manual, as required to effect all the revisions needed as a result of the conference.

As he proceeds the checker needs to be sure that he does not miss anything which has to be done as a result of the conference. To ensure this, while he is reviewing the answers recorded in the open spaces in the question sections, the conference notes, the conference print notations, and any other marks on the conference prints, he applies a light-green pencil tone over all the red conference notes and conference print notations, all pencil sketches and notes, and all other marks made at the conference. This light-green pencil tone is a graphic indication of the checker's progress in reviewing the various notes, and helps to keep him from missing even the smallest note.

As the checker takes each question and its recorded answer into account, and takes the actions necessary in accord with that question and answer, he should draw a vertical green line through the question and its answer. Colorbrite Dark Green #2128 is an appropriate green for this purpose. (This vertical green line is drawn in the same way as described in Article 8-2, and as shown in Plate III.) This green line should be drawn right beside the vertical red line (see Article 15-12) which indicates that a question has been presented and answered. Segments of this vertical green line (each segment being drawn through a different question and answer) connect and form an unbroken green line from top to bottom of the page when all of the questions and answers have been taken into account and require no further attention.

As with the questions and answers, the checker applies a vertical green line on the conference notes to indicate that they have been taken into account and require no further attention.

After the checker has completed his review of the questions and answers, conference notes, etc.; has made the necessary checker's notations on the master check-prints; and has entered the necessary related comments and sketches in the checker's manual—he transmits these notations, comments, and sketches to the persons responsible for seeing that the revisions are made. (Transmittal of these notations, comments, etc. is described in Chapter 11.)

In addition to seeing that the drawings are revised in accordance with the conference, the checker takes any other actions required as a result of the conference. This includes recording in the checker's manual any additional questions which might have arisen at the conference and communicating with others regarding the conference.

15-23 CONFERENCE NOTES FILE

The checker should provide himself with a three-ring loose-leaf binder called the conference notes file, which is used as the file for all of the conference notes made by the checker at conferences.

After each conference, the checker staples the pages of conference notes from the conference into a group and files the group in the conference notes file. The stapled groups should be arranged in reverse chronological order, with the latest group on top.

Remark: A variation on this basic method of filing all groups of conference notes in reverse chronological order is recommended for projects of an appreciable size. When conference notes relate to a single division of the work (or to a single individual), the checker files those conference notes together (in groups and in reverse chronological order). For example, all conference notes resulting from conferences with structural engineering personnel are filed together, and the same applies to HVAC, plumbing, and electrical conferences. When a conference involves personnel from more than one division of the work, the checker should file those conference notes under a general conference notes grouping, rather than under a particular division of the work.

Conference notes from conferences with a single individual are grouped together in the conference notes file and are treated in the same way as the notes for a division of the work. If an organization is involved, the conference notes related to various individuals in the same organization would be grouped together.

15-24 INACTIVE CONFERENCE NOTES FILE

On large projects, the checker should provide himself with an additional three-ring loose-leaf binder which is called the **inactive conference notes file**. This binder is used to file all inactive conference notes; inactive meaning that the checker has taken the actions required to carry out the intent of the notes.

Since the conference notes are stapled into a separate group of pages for each conference, each of these groups should remain intact and should not be broken up to retire a single page which has reached inactive status. Instead, after all the conference notes in a group of pages have been acted on by the checker, all the pages in that group are retired at once, without taking the group apart. As long as even one conference note in a group of pages has not been taken into account, that entire group of pages should remain in the active conference notes file.

15-25 DUPLICATE CONFERENCE PRINT NOTATION NUMBERS

If possible, the checker should not have another conference until he has reviewed the answers, conference notes, and conference print notations made at the preceding conference, and has applied light-green pencil over the notes and notations made there (see Article 15-22). The reason for this is that if he uses the same conference prints, the numbers of the conference print notations from the first conference would become confused with the numbers from the second conference, since these numbers normally start over with each conference. To prevent this problem, if one conference must closely follow another, the checker should run the numbers of the conference notes and their related conference print notations continuously through the two conferences.

Remark: Actually, it would be an excellent practice if numbers of conference notes could be run continuously through *all* conferences. The problem is that it is hard to be sure at each conference exactly what the last conference note number was unless notes from the previous conference are available.

Another solution to the problem of confusing the conference print notation numbers placed on the prints at different conferences is for the checker to include the date of the conference with each conference print notation. This date could be added to each conference print notation after the conference, to avoid slowing the checker down during the conference. If more than one conference is held on the same date, the checker could add a subscript ($_{A, B, C,}$ etc.) to the date in order to distinguish different conferences.

15-26 WRITTEN CONFERENCE REPORT

If the checker needs to prepare a written conference report, he would be aided by the recorded answers to his questions, as well as by his conference notes and sketches.

15-27 CONFERENCES CONDUCTED BY PROJECT ARCHITECTS, PROJECT MANAGERS, AND OTHERS

The techniques described for checker's conferences apply to any conferences which involve drawings. Such conferences include those conducted by project architects, project managers, job captains, and designers. No matter who uses them, the techniques described in the preceding articles aid in conducting conferences, keeping conference notes, and marking the drawings in an orderly and efficient manner.

16 CHECK-PRINT OVERLAYS

16.1 CHECK-PRINT OVERLAYS

Another basic concept of this system of checking is a technique for checking a print when the drawing is incomplete and further checking on more complete prints will eventually be required. Ordinarily, any checking done on an incomplete print must be repeated each time a more complete print is issued; this problem is overcome by the overlay technique.

In this technique, a sheet of thin, highly transparent, yet reasonably tough and stable tracing paper is stapled over the master check-print. The checker then applies green and yellow pencil tones and makes desired notations in red pencil using the method described in Chapter 3, *except* that the colored pencil is applied on this tracing paper overlay rather than directly on the master check-print. (The tracing paper overlay is called the **check-print overlay** because each overlay is placed on top of a master check-print.)

The advantage of using this check-print overlay is that it may be removed from the master check-print onto which it is stapled and be transferred—*along with all the checker's indications in colored pencil made up to that point*—to a later, more complete master check-print. In this way, the checker's indications in colored pencil may be transferred almost instantly, without having to spend hours forwarding the indications to each successive, more complete master check-print or having to completely recheck each new check-print.

16-2 MATERIAL FOR CHECK-PRINT OVERLAYS

A tracing paper which has the proper characteristics for use with the check-print overlay technique is Aquabee #525, made by the Bee Paper Company. Aquabee #525 is very transparent (which is absolutely essential for this use), yet it is tough and stable enough to withstand the requisite marking and handling, as well as the repeated stapling, removal, and restapling involved in transferring the overlays from one master check-print to another. The Aquabee #525 tracing paper has an excellent surface for the recommended smooth application of the colored pencil, and the colored pencil may be readily erased where necessary. (An effective eraser for this purpose is the Faber-Castell #1962 Auto-Magic Rub vinyl eraser.)

Other manufacturers' products which have the same transparency, toughness, and a suitable surface might be used. Some of these are supplied in pads or cut sheets of the desired size, which saves considerable time over having to cut the sheets from a roll.

16-3 ALTERNATE POLYESTER OVERLAYS

Polyester film, in 2- or 3-mil thicknesses with a matte surface on one side, may be used for check-print overlays on large projects where hard usage is expected. Most films can be cleaned for reuse with rubbing alcohol; if rubbing alcohol damages a particular film, water and detergent may usually be safely used instead.

16-4 ATTACHING CHECK-PRINT OVERLAYS

Each check-print overlay should be the same size as the master check-print to which it will be attached. If cut sheets or pads of tracing paper are unavailable in the right size, the sheets should be cut (or torn, using a parallel rule as a straightedge) from a suitable roll of tracing paper, as described above.

Each properly sized check-print overlay should be aligned to fit squarely over the master check-print to which it will be attached, and it should be attached with four staples, each placed about 1 to 2 in. from the corner of the sheet.

Before the staples are applied, a piece of drafting tape, approximately ¾ × 1½ in., should be placed on the check-print overlay at the four points to be stapled; this tape serves to reinforce the tracing paper so that it can withstand the wear and tear of being transferred from one master check-print to another.

16-5 DATE AND SHEET NUMBER ON CHECK-PRINT OVERLAY

Prior to attaching a check-print overlay onto a master check-print, the checker should indicate on the check-print (not on the overlay) the date, proper check-print designation, and issue-number in Colorbrite Green #2128 (as described in Chapter 2).

After the checker attaches a check-print overlay onto the initial master check-print (see Article 16-4), he should place the sheet number of the master check-print on the check-print overlay. (If the master check-print does not have a sheet number, the checker should assign it a temporary sheet number and place that same temporary number on the check-print overlay.) The sheet number should be placed in a convenient position near the lower-right corner of the check-print overlay.

In addition to the sheet number, the checker should indicate the date on which the overlay was attached to the initial master check-print. Also, registration marks should be placed on the check-print overlay in order to transfer it to later master check-prints. (The application of these registration marks will be described in Article 16-6.) The checker should use Colorbrite Green #2128 for the sheet numbers, dates, and registration marks he places on the check-print overlay.

16-6 TRANSFERRING CHECK-PRINT OVERLAYS

Check-print overlays may be transferred from one master check-print to another by removing the staples from the overlay and then restapling it onto the later master check-print. In this transfer, the check-print overlay must be restapled in exactly the same position on the later master check-print as it was on the earlier master check-print. This is done by means of the registration marks that the checker places on the check-print overlay.

When the checker attaches the check-print overlay to the initial master check-print, he should align it to fit squarely over the master check-print, and then staple it in place in that position. After stapling the check-print overlay in place, the checker should place registration marks on the check-print overlay over the lower-left and lower-right corners of the border of the master check-print, using Colorbrite Green #2128 for the registration marks.

Then, when the check-print overlay is removed from the initial master check-print, it may be relocated in exactly the same position on the next master check-print by placing these registration marks precisely over the lower-left and lower-right corners of the border of the new master check-print and carefully stapling the check-print overlay in that position.

16-7 APPLICATION OF CHECKER'S COLOR CODE ON CHECK-PRINT OVERLAYS

The light-green and yellow pencil tones and the red pencil notations of the checker's color code are placed on the overlay in the same manner they would be placed on the master check-print.

Visually, the effect of applying the colored pencils on the check-print overlay is practically the same as if the colored pencils were applied directly to the master check-print. This is due basically to the transparency of the tracing paper; in addition, the light-green and the yellow pencil tones are transparent enough, if properly applied, that items on the master check-print below are still adequately visible after the tones have been applied to the overlay.

16-8 ITEMS PREVIOUSLY CHECKED, ITEMS NOT PREVIOUSLY CHECKED, AND ITEMS ADDED

When the check-print overlay is transferred and stapled in place over a later master check-print, the checker is able to determine which items have been previously checked, and which items have not—including items added to the drawings after the earlier print was made, which thus appear on the later master check-print but not on the earlier one.

The checker can do this because the light-green pencil tones previously applied to the check-print overlay show which items have already been checked; conversely, the absence of green indicates which items on the new check-print have not been checked. Also, where new items have been added, it is immediately apparent that they have not been checked because no green tone appears on the check-print overlay for such added items.

16-9 ITEMS REMOVED FROM THE DRAWINGS

Besides showing where added items on a new check-print have not been checked, the check-print overlay also shows where items which previously appeared on a print (and had been checked and had green applied to them) have been removed from the drawing. This may be readily discerned by the fact that an area on the check-print overlay with light-green pencil tones appears, but there is no matching item on the check-print below. If a different item has been added to the check-print where the previous item has been erased, the new item usually will not match the pattern of the light-green pencil tones on the check-print overlay.

Where items have been removed from the drawings and the light-green tone has been left on the check-print overlay, it should be erased since it is no longer applicable. If the area of green involved is large, it should be cut out of the check-print overlay and a new, matching piece of tracing paper should be spliced in with transparent mending tape. The erased area on the check-print overlay (or the spliced-in area of new tracing paper) is then used by the checker in the same manner as any other part of the check-print overlay.

16-10 COMMUNICATING WITH OTHERS BY USE OF CHECK-PRINT OVERLAYS

In communicating with others, the check-print overlays are used exactly as if the checker had placed his notations directly on the check-prints themselves. The checker copies his notations onto duplicate check-prints; applies yellow over the red notations on the overlay to indicate that they have been copied; and sends these duplicate check-prints, along with copies of the related comments, sketches, etc., to those with whom the checker wishes to communicate (see Chapter 11).

16-11 VERIFYING REVISIONS BY USE OF CHECK-PRINT OVERLAYS

After a drawing has been revised in accord with the checker's communication with others, the checker then verifies that the revisions have been made correctly. This verification is generally as described in Chapter 13, and consists of checking a new check-print (of the revised drawing) against the yellow/red notations on the check-print overlay, which would still be attached to the earlier check-print. The verification is done at the same time as the related comments, sketches, etc., are checked in the checker's manual.

When the checker verifies that a revision has been properly made, he then *erases* the yellow/red notation (related to that revision) from the check-print overlay, thus indicating that the notation has been taken into account and requires no further attention. (This erasure of inactive notations from check-print overlays is different from previously described procedures and is discussed further in the following article.)

If any revisions have been improperly made, or not made at all, the checker would not erase the related notation on the check-print overlay, but would use it (and any necessary new notations) as the basis for further communications with others as required to see that the revision is eventually made correctly.

16-12 ERASING INACTIVE NOTATIONS FROM CHECK-PRINT OVERLAYS

As described above, when the checker has verified that a revision has been properly made, he then *erases* the yellow/red notation (related to that revision) from the check-print overlay, thus indicating that the notation has been taken into account and requires no further attention.

If the yellow/red notation had been made directly on a check-print (rather than on an overlay), the checker would have applied a light-green pencil tone over the notation that was no longer needed, instead of erasing it. The reason the checker erases the inactive yellow/red notation from the check-print overlay rather than applying light green over it is that when he transfers the check-print overlay to a later print, if the notation is not erased the inactive yellow/red notation would interfere with the visibility of the new print below the overlay. If the inactive notation is erased, the checker then would have a clear view of the new print below and could proceed with his checking without interference.

16-13 ADVANTAGES OF THE CHECK-PRINT OVERLAY TECHNIQUE

The check-print overlay technique allows the checker to transfer his indications in colored pencil from early, incomplete check-prints to successively more complete check-prints without having to forward all of those indications to each of the later check-prints.

This ability to check early, incomplete check-prints and then check successively more and more complete check-prints, in an efficient manner without having to start over on each new check-print, gives the checker the advantage of being able to begin checking at an early stage. No longer must the checker wait until the drawings are complete before he begins checking. The ability to start checking early prevents the situation in which the checker is faced with the necessity of performing all of the checking—plus having to get the revisions made and then verified—all at the last minute in a time span which is often very inadequate.

In addition, there is obviously considerable value in allowing the checker to bring his expertise to bear during the early stages of drawing production; early involvement allows the checker to aid the project manager and job captain in detecting problems, which are much easier to correct early than they are later on.

16-14 USING CHECK-PRINT OVERLAYS

Throughout the previous chapters, the descriptions of the checker's procedures have been based on the marking of his color-code tones and notations directly on the master check-prints. If the check-print overlays are used, all descriptions of procedures would still be completely applicable; the only difference being that the checker would be marking his color-code tones and notations on the check-print overlays instead of directly on the master check-prints.

It should be emphasized that if the checker is working with check-prints which are incomplete to almost any degree, he should employ the check-print overlay technique for marking his color-code tones and notations.

Remark: One minor difference in using check-print overlays should be noted: while it is suggested in earlier chapters that the master check-prints be folded to a convenient size for cross-checking against other master check-prints (which might also be folded), the checker will find that it is preferable *not* to fold master check-prints which have check-print overlays attached over them. Folding would impair the close registration needed between the color-code tones and notations on the overlay above and the related items on the master check-print below.

16-15 FINAL CHECK OF COMPLETE SET OF CHECK-PRINTS

Without detracting from the value of using the check-print overlay technique, which is an indispensable aid in checking drawings during the ongoing production process, there is still some chance that something will be changed without the checker's knowledge, and that he will fail to detect the change even with the use of the check-print overlays. For that reason, even though the drawings have been continually checked (using the check-print overlay technique) during the production process, it would still be advisable to completely recheck a complete set of check-prints when the drawings are finished. This might even be done after the drawings have been issued to the contractors to prepare bids (with any changes being covered by addenda or by issuing revised drawings).

If the checker is given the opportunity to recheck a complete set of finished drawings, he should do so. He would not need to use the check-print overlay technique, but instead he could mark the color-code tones and notations directly on the check-prints of that complete set. The time required for such a complete rechecking should not be too great because the checking done (using the check-print overlays) during the production process should ensure that there will be a minimum of problems that would require extensive and time-consuming comments and sketches.

16-16 CHECKING A TRACING WITH THE OVERLAY TECHNIQUE

Occasionally it is advantageous to be able to check a tracing without making a check-print. This is easily done by use of basically the same overlay process as described in the preceding articles, with certain modifications as described in the following paragraphs.

To check a tracing with the overlay technique, the checker first tapes the tracing onto a drawing board. Over the tracing, the checker tapes a sheet of clear acetate, in a 5- or 7.5-mil thickness. (The acetate sheet is needed to protect the tracing from possible damage, or from any deleterious effect on the drawing lines if they are made in pencil.) Over the acetate, the checker tapes a thin tracing paper overlay as described in Article 16-2, or a polyester overlay as described in Article 16-3. The checker then proceeds to check the underlying tracing, applying green pencil tones and red pencil notations on the overlay, in the same manner as described for overlays on check-prints.

After he has finished checking all items on the tracing below the overlay (and has marked the overlay with needed notations in red), the checker (or a drafter) reverses the positions of the tracing and the overlay. Thus, he tapes the overlay on the drawing board, the acetate over the overlay, and the tracing over the acetate. The checker (or drafter) then proceeds to make revisions of the tracing, using the underlying notations in red to guide him.

After he has made the revisions, the checker or drafter again reverses the positions of the tracing and the overlay, with the tracing taped on the drawing board, the acetate over that, and the overlay taped on top of the acetate. Now the checker proceeds to verify that the revisions have been made correctly. He takes each red notation and ascertains that the tracing has been revised in accord with the notation. If it appears that the revision has been correctly made, the checker would erase the red notation from the overlay, and would then apply light-green pencil tone over elements of the revision (double-checking to see that they are correct). This process is repeated for each red notation, until all of them have been erased, and all revisions have had a light-green tone applied over them.

17 FLOOR PLAN OVERLAYS

17-1 FLOOR PLAN OVERLAYS

This procedure consists of a graphic technique which helps the checker coordinate the floor plans of the architectural and the engineering divisions of the work. It utilizes a tracing paper overlay technique similar to the one described in Chapter 16. The way a **floor plan overlay** is prepared and used is described in the following paragraphs. (See also Plate IV, which illustrates the following procedures.)

First, the checker obtains a print of the architectural floor plan. The checker then tapes this print on a drawing board, and an overlay of thin tracing paper is taped on top of it, as described in Chapter 16. The checker indicates a date, title (floor plan overlay), and identification of the floor involved (e.g., first floor). He also provides registration marks on the overlay, *over widely separated corners of the architectural floor plan below.* (The date, title, floor number, and registration marks should be in Colorbrite Green #2128, as described in Chapter 16.)

The checker then applies Colorbrite Orange #2122 to the overlay, over the following: configurations of walls and partitions; indications of windows, doors, and door swings; and outlines of fixtures, fixed equipment, and other floor plan elements. This orange color should be carefully applied on the overlay, either freehand or with a parallel edge and triangles. The color should be smooth, even, and transparent.

Next, the checker obtains a print of the HVAC floor plan; he tapes this print on the drawing board and tapes the floor plan overlay over the print. Then the checker proceeds to carefully examine all areas (one space at a time) for any conflicts revealed by differences between the orange tone on the overlay and the print below.

The transparent orange allows enough of the underlying blueline print to show through that the checker can discern where the overlying orange tone and the print are in agreement—indicating that the print of the HVAC floor plan is in agreement with the latest architectural floor plan.

On the other hand, a conflict would be indicated in two ways: at any point where there is orange on the overlay with no corresponding indication on the print below, and at any point where there was some indication on the print below with no corresponding orange tone on the overlay above.

If the checker detects any such conflicts, he would mark the floor plan overlay with checker's notations (in red pencil) and would enter any comments or questions in the checker's manual, to ensure that the conflicts will be resolved. (Plumbing and electrical floor plans are coordinated in the same manner as HVAC plans, as described above.)

17-2 USING FLOOR PLAN OVERLAYS

Floor plan overlays, marked with checker's notations, essentially become check-print overlays and are stapled over architectural floor plan check-prints. As such, the floor plan overlays are included with the master check-prints.

The checker uses the floor plan overlays in the same way as check-print overlays are used—to communicate any problems to the proper HVAC, plumbing, and electrical personnel, and also to verify the correctness of any revisions when they have been made. The communication of problems would be as described in Article 16-10, and the verification of the revisions would be as described in Article 16-11.

17-3 ALTERNATE POLYESTER OVERLAYS

If a project is large, the checker might elect to use polyester film (in lieu of the thin tracing paper) for the floor plan overlays, in the same manner as described in Article 16-3.

18 CEILING PLAN OVERLAYS

18-1 CEILING PLAN OVERLAYS

Another concept of this system of checking is a graphic technique for ensuring that all the diverse features which appear in the ceiling of a building are coordinated with one another, and are properly shown on the architectural ceiling plan.

This procedure utilizes a tracing paper overlay technique similar to the one described in Chapter 16. The way a **ceiling plan overlay** is prepared and used is described in the following paragraphs. (See also Plate V, which illustrates these procedures.)

First, the checker acquires a print of the architectural ceiling plan. He tapes this print on a drawing board and tapes an overlay of thin tracing paper on top of it, as described in Chapter 16. The checker indicates a date, title (ceiling plan overlay), and identification of the floor involved (e.g., first floor). He also provides registration marks on the overlay, *over widely separated corners of the architectural ceiling plan below.* These registration marks are used later to allow the ceiling plan overlay to be properly placed over HVAC and electrical plans. (The date, title, floor number, and registration marks should be in Colorbrite Green #2128, as described in Chapter 16.)

Next, the checker removes the ceiling plan overlay from the board and tapes it over a print of the HVAC plan, using the registration marks to properly locate the overlay on top of the HVAC plan. He then applies Colorbrite Orange #2122 on the overlay, over all ceiling diffusers, grilles, and other HVAC features which appear in the ceiling. This orange tone should be applied very lightly and smoothly. Next, the checker tapes the ceiling plan overlay over a print of the electrical lighting plan. He applies a yellow tone on the overlay, over all lights and other electrical features which appear in the ceiling. Again, this tone should be applied very lightly and smoothly.

At this stage, the orange tone placed over all HVAC items in the ceiling and the yellow tone placed over all the electrical items in the ceiling serve to indicate any conflicts between items appearing in the ceiling as shown on the HVAC and electrical drawings. The checker indicates any conflicts by marking notations on the overlay, using Colorbrite Medium Red #2126.

Next, the checker *staples* the ceiling plan overlay over a print of the architectural ceiling plan. He applies a light-green tone on the overlay, over all orange and yellow tones which have corresponding HVAC or electrical indications on the architectural ceiling plan below them. This light-green tone indicates that the architectural ceiling plan below is in agreement with the HVAC and electrical plans at those points.

If the checker observes an item on the architectural ceiling plan below and there is no orange or yellow tone on the overlay above, he would draw an outline in red pencil on the overlay, tracing over the outline of the item on the print below. This red pencil outline on the overlay signals that a problem exists. If the checker observes either an orange tone or a yellow tone on the overlay and there is no corresponding item shown on the ceiling plan below, he would draw an outline in red pencil on the overlay around the orange or yellow tone. Again, this red pencil outline on the overlay signals that a problem exists.

In addition to using the red outline on the overlay to indicate problems, the checker also makes any desired checker's notations in red pencil on the overlay, along with related comments in appropriate sections of the checker's manual, for use in communicating problems to other persons.

18-2 USING CEILING PLAN OVERLAYS

Ceiling plan overlays, marked with red pencil where problems occur, essentially become check-print overlays (stapled over check-prints of the architectural ceiling plans), and as such they are included with the master check-prints.

The checker uses the ceiling plan overlays in the same way as he uses check-print overlays—to communicate any problems to the proper architectural, HVAC, or electrical personnel, and also to verify the correctness of any revisions. (The communication of problems is described in Article 16-10, and the verification of the revisions is described in Article 16-11.)

18-3 ALTERNATE POLYESTER OVERLAYS

If a project is large, the checker might elect to use polyester film (in lieu of the thin tracing paper) for the ceiling plan overlays, as described in Article 16-3.

19 ARCHITECTURAL AND ENGINEERING OVERLAYS

19-1 ARCHITECTURAL AND ENGINEERING OVERLAYS

Another concept of this system of checking is a graphic technique for simultaneously coordinating elements of architectural, structural, HVAC, plumbing, and electrical drawings.

This concept utilizes a tracing paper overlay technique similar to the one described in Chapter 16. The way in which an **architectural and engineering overlay** is prepared and used is described in the following paragraphs. (See also Plate VI, which illustrates these procedures.)

First, the checker acquires a print of the HVAC plan; then he tapes this print on a drawing board, and over it he tapes an overlay of thin tracing paper. The checker indicates a date, title (A & E overlay), and identification of the floor involved (e.g., first floor). He also provides registration marks on the overlay, *over widely separated corners of the HVAC plan below.* These registration marks are used later to allow the A & E overlay to be properly placed over architectural, structural, plumbing, and electrical plans. (The date, title, floor number, and registration marks are in Colorbrite Green #2128.)

Next, the checker removes the A & E overlay from the board and tapes it over a print of the architectural floor plan, using the registration marks previously provided to properly locate the overlay on top of the architectural floor plan. The checker then applies Prismacolor Copenhagen Blue #906 on the overlay over all walls and partitions, omitting the color at openings such as doors, windows, louvers, etc. Next, the checker tapes the A & E overlay over a print of the structural framing plan. He then uses Colorbrite Scarlet #2166 to trace the structural framing. The checker next tapes the A & E overlay over a print of the plumbing plan showing piping at the ceiling. He uses Colorbrite Green #2128 to trace the piping at the ceiling. After this step, the checker tapes the A & E overlay over a print of the electrical lighting plan. He then applies a yellow tone over all lights or other electrical items which appear in the ceiling. And last, the checker *staples* the A & E overlay over a print of the HVAC plan, showing the ducts, diffusers, and grilles above.

The checker now selects a convenient point and begins to apply Eagle Turquoise Green (light green) on the overlay, over a duct as shown on the HVAC plan below the overlay. As he proceeds to apply this light-green tone, the checker indicates (in Colorbrite Red #2126) a section ① cut at the first point which appears to be critical in some way. This section cut should appear as shown in Figure 19-1.

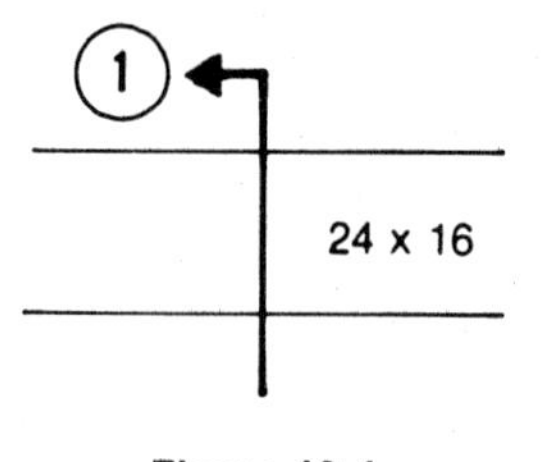

Figure 19-1

Then the checker takes an 8½ × 11 in. note pad and draws a small schematic section, freehand, to the desired scale. (This schematic section should be drawn with soft graphite pencil.) The checker refers to the structural drawings to get information regarding the size of the structural framing at that point, and to the architectural drawings to get information regarding the ceiling at that point. If any lighting is shown at that point, as indicated by the yellow tone applied on the overlay previously, the checker refers to the electrical drawings to get information regarding lighting fixtures (such as whether the fixture is mounted flush in the ceiling, is surface-mounted, or is on stems, and other critical information). If any piping is in the area, as indicated by Colorbrite Green #2128 applied on the overlay previously, the checker refers to the plumbing drawings to get information on pipe sizes and heights above the floor at that point, and any other critical information.

As he thus reviews the structural, architectural, electrical, and plumbing drawings, the checker indicates on the schematic sketch all critical conditions he encountered at the point where this section ① is located. Figure 19-2 is an example of how section ① might appear.

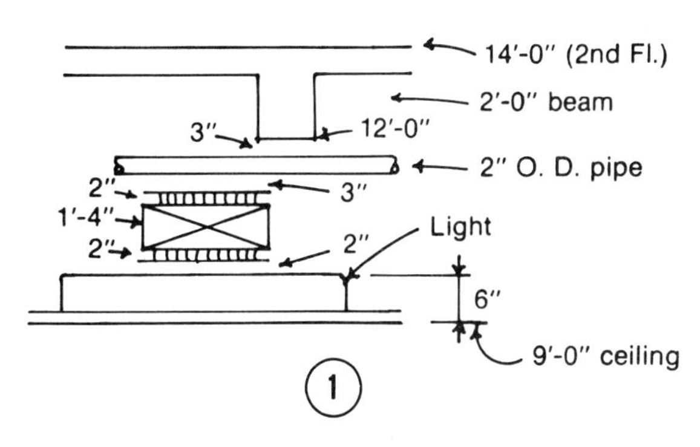

Figure 19-2

In Figure 19-2, the dimensions of various elements indicate a 9-ft ceiling height. If the architectural drawings indicate a 9-ft ceiling height, it would be verified by this section ①.

Assuming that everything is in order at section ①, the checker proceeds to extend the light-green tone on the overlay, over the indication of the duct in the area of section ①. Also in the area of section ①, the checker applies a tone of Prismacolor Raw Umber #941 on the overlay, over the indication of the structural beam, as outlined by the Colorbrite Scarlet #2166 applied on the overlay previously.

The checker continues this application of light green over the HVAC ducts and raw umber over structural beams, until he gets to some point where conditions change. Such a change of conditions might be in the size of the duct, in the structure, in the ceiling height, or in any other area which would affect the relationship of the various elements of construction.

Where such a change of conditions occurs, the checker indicates another section cut, section ②, and then proceeds to draw another freehand schematic section on the 8½ × 11 in. note pad. The checker then repeats the process described above, drawing all critical conditions at that point on the new schematic section and checking to see that there are no conflicts.

All pages of schematic sections (or other sketches) are numbered, dated, and filed in a three-ring binder for future reference. The pages are filed in reverse chronological order, with the latest page on top.

In addition to checking for conflicts between the HVAC ducts, structural beams, piping, lighting fixtures, and architectural ceilings, the checker also checks along the route of the HVAC ducts and the beams for any places where the ducts and beams penetrate (or go over) architectural partitions or other construction features. At such points, the checker draws the same type of schematic sections (or other sketches such as schematic plans, elevations, etc.) to ascertain whether there are any conflicts.

In this manner, the checker progresses along the various ducts and beams, drawing schematic sections (or other sketches) at critical points and applying light green over the HVAC ducts and raw umber over the structural beams.

The checker never applies light green over the HVAC ducts, nor raw umber over the structural beams, beyond the point where he has analyzed all critical conditions and is satisfied that no conflicts exist. The checker thus has a graphic record of his progress, so that if he stops at any point, he knows exactly where to resume checking.

As he progresses, in addition to applying the color tones over the HVAC ducts and the structural beams, the checker also applies light green on the overlay over all room names, notes, equipment, symbols, and other items on the HVAC plan below. If any of these items affect the coordination of various elements in any way, the checker would analyze the situation for conflicts.

If the checker needs to note any miscellaneous memoranda on the overlay, such as a beam size or a critical dimension, he should make such notes in either graphite pencil or Colorbrite Green #2128. Light green may be applied later over the notes in graphite pencil to indicate that no further action is called for.

As the checker progresses, he notes any conflicts or other problems by indicating numerical notations in Colorbrite Medium Red #2126 on the overlay. As he makes these red numerical notations, the checker makes a like-numbered entry on a second 8½ × 11 in. note pad, which either describes any problems or illustrates them by means of a sketch. (These entries would be similar to notes and sketches made at conferences, as described in Chapter 15.)

All pages of these notes and sketches are numbered, dated, and filed in a three-ring binder for future reference. The pages are filed in reverse chronological order, with the latest page on top.

The checker continues until he has gone completely over the HVAC plan; has applied light green on the overlay over all HVAC ducts; has applied raw umber over the indication of all beams; has drawn all necessary schematic sections (or other sketches); has applied light green over all room names and other miscellaneous indications on the HVAC plan; has made all necessary notations in red on the overlay; and has recorded all related notes and sketches.

19-2 USING ARCHITECTURAL AND ENGINEERING OVERLAYS

Architectural and engineering overlays, with notations regarding any problems marked in red, become essentially check-print overlays, and are stapled over prints of the HVAC plans. Since A & E overlays have information from architectural, structural, HVAC, plumbing, and electrical drawings, they are filed in a separate group related to the master check-prints, which are grouped according to divisions of the work.

The checker uses the A & E overlays in the same way as he uses check-print overlays—to communicate any problems to the proper architectural, structural, HVAC, plumbing, or electrical personnel, and also to verify the correctness of any revisions. (This communication of problems is described in Article 16-10, and the verification of the revisions is described in Article 16-11.)

19-3 ALTERNATE POLYESTER OVERLAYS

If a project is large, the checker might elect to use polyester film, in lieu of the thin tracing paper, for the A & E overlays, as described in Article 16-3.

19-4 ADAPTING PROCEDURES TO OTHER SYSTEMS

It would obviously be impossible to give examples of A & E overlays for all of the diverse structural, HVAC, plumbing, electrical, and other systems which might be used. The example shown in Plate VI and described in this chapter would have to be adapted as necessary for use with other systems, but the basic principles remain the same.

20 APPLICATION OF CONCEPTS

20-1 GENERAL

The following is a step-by-step guide to the application of the concepts described in previous chapters. The sequences described are for a hypothetical full set of completed architectural and engineering working drawings. In these descriptions, the checker makes his checker's notations and indicates the checker's color code in colored pencil directly on the check-prints.

If the drawings were in earlier stages, the checker would use check-print overlays over the check-prints, and would mark checker's notations and checker's color code in colored pencil on the overlays, as described in Chapter 16. In the earlier stages, drawings obviously are not checked at points where items or sheets are missing, but other than that, the procedures described in the following guide are basically the same, whether the check-print overlays are employed or not.

Unless he is advised otherwise, the checker proceeds on the assumption that the working drawings he receives properly reflect all requirements established in the previous schematic design phase and in the design development phase. This means that all code and zoning requirements have been met, and so have all previously established requirements regarding the physical elements of the project, such as the type of structural system, the size and arrangement of spaces, the materials and finishes used, and the type of mechanical and electrical systems to serve the project.

The checker, of course, endeavors to the best of his ability to detect code violations, to question dubious use of materials or methods of construction, and to question any item on the drawings which seems to be unclear. The checker should not, however, be expected to go back to the beginning of a project and repeat all of the code research and other work required in the earlier stages of the project development. If this is expected of the checker, he must be so advised and be furnished with the necessary criteria, such as approved preliminary drawings, etc.—and, most important, he must be given the time to perform this additional checking.

In checking the work of the consulting engineers, the checker is expected to check only the coordination of the physical elements of the engineering, such as conflicts of ducts and beams, etc. Such things as the sizing of structural members, reinforcing, etc.; the sizing of HVAC equipment, ducts, etc.; and the sizing of electrical equipment, wiring, etc., are beyond the scope of the checker's responsibilities. The checker simply accepts such items as he finds them, with the understanding that the checking of all such items is the responsibility of the consulting engineers.

As far as is possible, the checker's analysis of all items of the construction should supplement the work of the project manager and job captains in ensuring that the desired features of the construction are adequately represented by the drawings. For instance, the checker should consider whether all conditions are shown by the wall sections. Here again, it must be understood that if the checker is expected to perform an extensive review of the project manager's and the job captain's work, adequate time must be allotted for him to perform these analyses thoroughly.

20-2 COMMUNICATION OF CHECKER'S COMMENTS, SKETCHES, AND NOTATIONS

Communication of the checker's comments, sketches, etc., to other personnel may be done either on a piecemeal basis during the checking or after the set is completely checked. In either case, communication should be done by transmitting duplicate copies of check-prints, comments, sketches, etc., as described in Chapter 11.

20-3 KEEPING THE CHECK-SET IN PROPER SEQUENCE

It is recommended that the check-set be kept in proper sequence at all times, using spring-type clamps to keep the sheets neatly aligned. These clamps should be applied, not to the usual left-hand binding edge, but to the upper border at the center. This allows easier access to all areas of each sheet. On large sets, use two clamps. On very large sets, subdivide the sets into manageable subsets, or substitute Plan-Hold friction binders for the spring clamps. These binders would have to be applied on the left-hand binding edge.

When extensive checking needs to be done on one sheet, it should be temporarily removed from the check-set. Likewise, when two or more sheets are to be cross-checked, they should be temporarily removed from the check-set.

20-4 CHECKER'S COLOR CODE

As the checker proceeds with the step-by-step procedures described in the following paragraphs, he applies light green to all items which he has checked and either has accepted tentatively as shown or has reviewed and accepted provisionally subject to actions indicated by checker's notations. This application of light green is described in Chapter 3.

In each step in the following descriptions in which it is indicated that the checker checks certain items, it is understood that the checker also applies the light-green pencil tone over the items after he has checked them. Also, any instruction to apply the light-green tone over items which have been checked will not necessarily be repeated in each step.

20-5 CHECKER'S COMMENTS, SKETCHES, QUESTIONS, AND NOTATIONS

In the following description of the step-by-step procedures used in checking a hypothetical set of drawings, it is understood that at any point in the procedures where the checker wishes to make a comment, draw a sketch, ask someone a question, or take any other action regarding the project, he takes such actions in the manner described in earlier chapters. This includes making checker's notations on the check-prints (see Chapter 4), and recording his comments, sketches, etc., in the checker's manual (see Chapters 5 through 8).

As stated, the checker may take these actions at any point in the procedure. This is not necessarily repeated at each step in the following paragraphs, as it would be repetitive and unwieldy.

20-6 STEP-BY-STEP PROCEDURES

Step 1 Checker is supplied with two sets of diazo blueline check-prints. The sets are not bound but are kept in proper sequence by use of one or more large spring clamps, applied to the upper border of the set (or by other means as described in Article 20-3).

Step 2 The checker places the designation MCP #1A (master check-print #1A) on each sheet of one set of check-prints, and DCP #1A (duplicate check-print #1A) on each sheet of the other set. The checker also provides a date on each sheet of the two check-sets. The designations MCP #1A and DCP #1A, and the dates on each sheet, are applied in Colorbrite Green #2128, as described in Chapter 2. (The designations MCP #1A and DCP #1A each include the 1A to indicate a new cycle of check-prints. The 1A differentiates these check-prints from any earlier issues.)

Step 3 The checker lays aside the set of prints designated DCP #1A for use later in communicating his comments, sketches, etc. The other set of prints, designated MCP #1A, is the set which the checker uses in checking. This set will be called the master check-prints, or more briefly check-prints, in the following descriptions.

Step 4 The checker takes the master check-prints, temporarily skips the cover sheet and the index sheet, and starts with the first sheet with drawings on it. As a typical example, assume that this first sheet is Sheet A-1, and that it shows a site plan and a number of exterior details.

The checker briefly examines each drawing item on the sheet and reviews each title and reference number. Thus, the checker briefly examines the site plan and reviews the title and reference number, which might be Site Plan 1/A-1. Observing that the title seems appropriate and that the reference number seems correct, the checker applies a transparent tone of light-green colored pencil over the indication of the title and the reference number. No colored pencil is applied to the site plan itself at this time.

The checker proceeds in this same way until he has reviewed all the items on this first sheet and has applied light green over each title and reference number if they appear to be correct.

As stated earlier, if the checker feels that he needs to record any comments or if he has any questions, he would make checker's notations in red pencil on the check-prints and would record his related comments, questions, etc., in the checker's manual. After making any desired checker's notations and recording related entries in the checker's manual, he then applies light green over the item in question (such as a title and reference number which he thinks is wrong). This light green does not indicate that the item is correct, but instead indicates that the item has been reviewed and is provisionally accepted—subject to the actions indicated by the red checker's notations and their related entries in the checker's manual. (The checker may make necessary checker's notations and record related entries in the checker's manual at any point in the step-by-step procedures described here and below.)

Step 5 The checker examines the sheet title of the first print, and if it appears to be correct, he would apply light green over the sheet title.

Step 6 The checker repeats Steps 4 and 5 on each sheet of the master check-prints.

Remark: The process of going through the entire set of master check-prints and briefly reviewing each drawing item and its title and reference number gives the checker a methodical overview of the project and of the drawings.

Step 7 The checker checks all section-cut indications on the architectural plans, building elevations, etc., against the sections themselves. If the section-cut indications are correct, he would apply light green to them. This process helps to initiate the checker into an understanding of how the building is put together, thereby enhancing his understanding of the plan of the building and the construction of the various elements.

Step 8 The checker takes a check-print of the architectural floor plan and prepares a floor plan overlay, as described in Chapter 17. The checker applies the floor plan overlay over each of the corresponding HVAC, plumbing, and electrical floor plan check-prints in turn, and checks for any conflicts, as described in Chapter 17. (If the project is a multistory building, the checker would then repeat this process with all the other architectural, HVAC, plumbing, and electrical floor plans.)

Step 9 The checker takes the check-print of the architectural floor plan, and, starting with the first-numbered space on the plan, examines the space name and number. If the space name seems appropriate and the number appears to be correct, he applies light green over the name and number.

Step 10 The checker checks for the name and number of that first space on the finish schedule, and if it is in agreement with the name and number shown on the plan, he would apply light green over the name and number on the schedule.

The checker proceeds to review the finishes scheduled for that space. In the absence of other criteria (such as finish schedules approved by the owner), the checker accepts the finishes as shown. He would, however, judge whether each of the finishes seemed suitable for that space, and if it did not, he would make a notation on the check-print and enter a comment or question in the checker's manual. For instance, if he saw that wood paneling had been indicated on the walls of an operating room, he would certainly want to check that with a job captain or consultant. As he reviews each finish indication on the schedule, the checker applies light-green pencil over the indication if it seems to be correct. If he has a comment, he would first make a notation in red on the check-print, would then enter his comment in the checker's manual, and would finally apply light green over the item in the schedule.

Step 11 The checker repeats Steps 9 and 10 for all spaces on all architectural floor plans and for all finish schedules. Cross-checking the space names and numbers on the plan against the finish schedule will show whether a space has been either omitted from the schedule or erroneously included.

Step 12 The checker finds the first-numbered door on the architectural floor plan and examines its location, direction of swing, etc. If he saw no problem, he would apply green pencil to the door's number on the plan. He then checks for that door number on the door schedule and applies light-green pencil over it. This cross-checking of the door number on the plan and the door number on the schedule shows whether a door has been either omitted from the schedule or erroneously included in the schedule.

The checker checks the door schedule. First, he compares the width as shown on the door schedule with the scaled width as shown on the floor plan. If these agree, and the width seems appropriate to him, the checker would apply light-green pencil over the width as shown on the schedule. Next, the checker examines the figures for the door height and thickness. If these seemed appropriate to him, he would apply light-green pencil over the figures, indicating that he accepts them subject to checking against later information from the door details and building elevations, etc. The checker does the same thing with the door material and frame material.

The checker removes from the set of master check-prints the sheets which contain the following: architectural floor plans, finish schedules, door schedules, and door details. The sheets should be folded, if necessary, and arranged as described in Article 10-7. The checker then proceeds to check the door details as follows.

Taking the first door on the door schedule, he checks each detail related to that door. First, he checks the door and frame materials as shown on the door details against the materials as shown on the door schedule. Next, the checker checks the finishes shown on the details of that door against the finishes shown on the finish schedule. Then, he checks the details of that door against the floor plan to see that wall thicknesses, placement of the door frame, and other elements agree. He checks all detail numbers on the details themselves and on the door schedules. As he proceeds with his checking, the checker applies light green on all items that he accepts. If he has any comments or questions, he would make the necessary notations on the check-prints and entries in the checker's manual.

After he has checked the door details of the first door on the door schedule, the checker takes each successive door on the door schedules and checks all the door details in the same manner, until all details of all doors have been checked.

After he has finished checking the door details, the checker replaces all the sheets which he has removed from the set of master check-prints.

Step 13 The checker removes the first architectural floor plan sheet from the master check-prints and proceeds to check it by using the single-sheet check technique described in Chapter 9.

The checker subjects each item of equipment, fixture, construction feature, etc., to the analysis described in Chapter 9. As he proceeds, he uses checker's notations to indicate other points on the drawings which should be cross-checked with regard to each item. He also uses the checker's notations, in conjunction with entries in the checker's manual, to indicate any comments and questions about any of the items.

The checker applies a light-green pencil tone over each item when he has completed the checking of an item. He continues thus until he has checked all items on the sheet.

Step 14 The checker checks all other architectural floor plans, using the single-sheet check technique.

Step 15 The checker checks each HVAC floor plan, using the single-sheet check technique.

Step 16 The checker checks each plumbing floor plan, using the single-sheet check technique.

Step 17 The checker checks each electrical floor plan, using the single-sheet check technique.

Step 18 The checker checks each structural framing plan, using the single-sheet check technique.

Step 19 The checker selects an architectural floor plan sheet and removes it from the set of master check-prints, along with the corresponding

HVAC floor plan sheet. The checker proceeds to cross-check the architectural floor plan against the HVAC floor plan, using the two-sheet cross-check technique.

Step 20 The checker takes the architectural floor plan selected in Step 19 and cross-checks it against the corresponding plumbing floor plan, using the two-sheet cross-check technique.

Step 21 The checker takes the architectural floor plan selected in Step 19 and cross-checks it against the corresponding electrical floor plan, using the two-sheet cross-check technique.

Step 22 The checker takes the architectural floor plan selected in Step 19 and cross-checks it against the corresponding structural floor framing plan, using the two-sheet cross-check technique.

Step 23 The checker takes all other architectural floor plans in turn and cross-checks each of them against the corresponding HVAC, plumbing, and electrical floor plans. He also checks them against the corresponding structural floor framing plans, using the two-sheet cross-check technique.

Step 24 The checker takes each HVAC floor plan in turn and cross-checks it against the corresponding plumbing, electrical, and structural floor framing plans.

Step 25 The checker takes each plumbing floor plan in turn and cross-checks it against the corresponding electrical and structural floor framing plans.

Step 26 The checker takes each electrical floor plan in turn and cross-checks it against the corresponding structural floor framing plan.

Step 27 The checker cross-checks the architectural ceiling plans against the corresponding HVAC and electrical lighting plans, using the ceiling plan overlays described in Chapter 18.

Step 28 The checker checks the relationship at all critical points of the structural members, HVAC ducts, piping, electrical lights, architectural partitions, ceiling construction elements, etc., using the architectural and engineering overlays described in Chapter 19.

Step 29 The checker cross-checks all architectural through-sections (sections through the entire building, usually at small scale) against the architectural floor plans and building elevations. In this cross-checking, the checker uses the multiple-sheet cross-check technique described in Article 10-7. The checker checks to see that through-sections, floor plans, and building elevations are in agreement as to materials, wall thicknesses, floor-to-floor heights, etc. He checks all notes on the through-sections. The checker considers whether enough through-sections have been shown on the drawings to give an adequate overall picture of the construction.

Step 30 The checker checks all large-scale wall sections against the architectural floor plans, building elevations, and through-sections. In this cross-checking, the checker uses the multiple-sheet cross-check technique described in Article 10-7. The checker checks to see that the sections, floor plans, and building elevations are in agreement as to materials, wall thicknesses, floor-to-floor heights, etc. He checks all notes on the sections. The checker considers whether enough large-scale wall sections have been shown on the drawings to give an adequate, detailed picture of the construction.

Step 31 The checker checks all architectural details (other than door details, which were checked in Step 12) against architectural floor plans, sections, and all other drawings affected by these details. This checking uses the multiple-sheet cross-check techniques described in Article 10-7.

Step 32 The checker checks the structural sections against the architectural sections for agreement of shapes, sizes, arrangement of members, etc., using the two-sheet cross-check technique described in Chapter 10.

Step 33 The checker checks all structural details against other structural drawings, architectural drawings, and other drawings affected by the structural details, using the techniques described in Chapter 10.

Step 34 The checker checks all HVAC details against the other HVAC drawings, architectural drawings, and other drawings affected by the HVAC details, using the techniques described in Chapter 10.

Step 35 The checker checks all plumbing details against the other plumbing drawings, architectural drawings, and other drawings affected by the plumbing details, using the techniques described in Chapter 10.

Step 36 The checker checks all electrical details against the other electrical drawings, architectural drawings, and other drawings affected by the electrical details, using the techniques described in Chapter 10.

Step 37 The checker checks all dimensions on the architectural drawings.

Step 38 The checker checks all dimensions on structural drawings. He cross-checks dimensions on the structural drawings against dimensions on the architectural drawings, using the two-sheet cross-check technique described in Chapter 10.

Step 39 The checker checks all other dimensions on all other drawings for the project. He cross-checks these dimensions against those on the architectural drawings, using the two-sheet cross-check technique described in Chapter 10.

Step 40 The checker checks the site grading plan against the topographical survey. He also reviews the site plan for positive drainage, inadequate slopes, excessive slopes, etc. If there is a separate site staking plan, the checker would check the dimensions on it and would cross-check it against the site grading plan.

Step 41 The checker cross-checks the building elevations against the architectural floor plans, using the two-sheet cross-check technique described in Chapter 10.

Step 42 The checker checks the building elevations against the site grading plan, checking the grades at the building.

Step 43 The checker checks the building elevations against structural, HVAC, plumbing, and electrical drawings for all features which should appear on the elevations (such as exposed structural elements, louvers, intakes, exhausts, etc.).

Step 44 The checker checks the roof plan. He cross-checks the roof plan against all architectural, structural, HVAC, plumbing, and electrical drawings for items which should appear on it.

Step 45 The checker reviews all drawing sheets in the set of master checkprints to see if any items remain which are still unchecked, indicated by the absence of a light-green tone applied over an item (or around its perimeter if the item is very large). The checker checks all unchecked items.

Step 46 The checker checks the index sheet and the cover sheet.

Remark: The actual checking would be complete at this point. The steps below concern communicating the checker's comments, sketches, etc.

Step 47 The checker communicates all of his comments, sketches, etc., to the persons who will be responsible for revising the drawings. The checker communicates this by use of the techniques described in Chapter 11.

Step 48 After the drawings are revised, the checker verifies the revisions, using the techniques described in Chapter 13.

INDEX

INDEX